D1432234

Cultural Uses of Plants

Cultural Uses of Plants

a guide to learning about ethnobotany

Gabriell DeBear Paye
Lead Teacher, Boston Public Schools

Edited by Elysa J. Hammond and Joy E. Runyon
Info-boxes developed by Elysa J. Hammond

The New York Botanical Garden Press

Published by
The New York Botanical Garden Press
Bronx, New York

The paper used in this publication meets the minimum requirements of the American National Standard for Information Sciences—Permanence of Paper for Printed Library materials, ANSI Z39.48-1984.

This book was printed in the United States of America
using recycled acid-free paper and soy-based ink.

Metropolitan Life Foundation is a leadership funder
of The New York Botanical Garden Press.

Contents

Preface

Why Teach Ethnobotany?

This curriculum was developed in response to my students' perennial question, "Why would I want learn about plants?" I found that ethnobotany, the study of how people use plants, is an excellent way to encourage students from diverse ethnic backgrounds to get excited about science. Asking students to learn about plants from the perspective of their own respective cultural backgrounds draws out students' natural inquisitiveness.

Cultural Uses of Plants challenges middle- and high-school students to design and implement their own unique experiments starting from a reference point they can relate to — a useful plant in their own culture. Using scientific methods, students carry out simple experiments to make discoveries about the plants they choose to study. Can mulberry bark be used to make paper? Does garlic work as an antibiotic? As their work unfolds, students learn to gather and record data, interpret results, and present their findings to others.

In 1987 I began developing this ethnobotany unit as an integral part of the curriculum for my biology and horticulture classes. Pursuing plant projects of their own design, many of my students have become winners of our school science fairs and have gone on to become winners in districtwide fairs for the City of Boston. Some of these and other projects are featured in this book.

When I taught botany using a more traditional textbook approach, I found the interest level and information retention rates among students much lower. In contrast, by entering the world of science through a personal experience in ethnobotany, students gain a real sense of how important plants are in their lives. I find that when they are fueled by this enthusiasm, they are more willing to then delve into other aspects of botany.

Good science education should empower students to participate in the discovery process rather than simply reading about the work of others or replicating that work almost mechanically. It should also touch their lives in a personal way. Using ethnobotany to teach science achieves both those goals. In spite of these merits, this topic remains relatively undiscovered by precollege science teachers. My objective in writing this book is to fill that need for a curriculum guide to ethnobotany for the middle and high school grades.

Overview

Each unit in this book begins with a brief article containing background information relevant to that unit's activities. At the end of each unit is a review section called Questions for Thought, followed by either Laboratory Activities or Field Explorations, and sometimes both. These sheets may be reproduced for class use.

Unit 1 introduces plants as the basis for our ecology and culture, defines the field of ethnobotany, and gives biographical sketches of three modern-day ethnobotanists. Unit 2 teaches the student how to conduct interviews for research and to gather background information in the literature. Unit 3 describes how to collect and preserve plants as herbarium specimens and how to propagate living plants; other methods of plant preservation are also presented. Units 4 and 5 describe simple lab procedures to test plants for basic nutritional and medicinal properties. Unit 6 offers several ideas for testing plants for other household or garden uses including paper making, dyes, perfumes, or ornamental use. Unit 7 guides students through the process of designing their own experiments, including the development and testing of hypotheses. Unit 8 presents the concepts of plant ecology and the importance of habitat conservation. Finally, Unit 9 offers suggestions for fun class activities to close this unit with a sense of celebration and an appreciation of the wider context of plant use in society; students will also find suggestions for careers and volunteer activities related to ethnobotany or to plants in general.

How to Use This Book

Teachers may choose to use the entire book as a whole unit on ethnobotany. In my classes it takes about one-third of a school year for a class to complete the entire curriculum (assuming a 180-day year and a 45–50-minute period five days a week). This unit lends itself easily to project-based learning, offering students the opportunity to create unique portfolio pieces that may serve them well later. At the end of this study they may have significant research results as well as unique materials to share with the class — everything from pressed plant specimens to botanical illustrations.

Teachers may also choose to integrate ethnobotany into related courses. As an interdisciplinary subject, ethnobotany draws upon many scientific fields including botany, anthropology, ecology, chemistry, medicine, history, and economics. Depending on the class being taught, different aspects of ethnobotany can be emphasized. In a biology class, the study of useful plants easily weaves into the unit on botany, and in an ecology unit the importance of biodiversity, habitat, and conservation can be emphasized. Microbiology is touched upon in Unit 5, where students discover how various plants affect microorganisms such as bacteria, protozoa, and yeast. In a health class the nutritional or medicinal value of plants may be highlighted by using Units 4 and 5. Math teachers will find information on statistical analysis and graphing for science projects in Unit 7. Language arts teachers and foreign language teachers can use Unit 2 to help students structure oral interviews and write papers about plants from another culture or their own, in English or another language. And in a social studies, history, or economics class, the stories of people and plant resources furnish a rich supply of teaching material (consider the link between sugar cane and the slave trade in the New World, the impact of the potato famine in Ireland, and the consequences of the Dust Bowl in the American Midwest).

When students obtain good experimental results, they find it rewarding to share these findings with others in the larger scientific community. At the end of most units in this book there is a summary sheet that students can also use as a means of sharing the results of their research. By emailing the author at gdpaye@hotmail.com, students and teachers can receive instructions on how to submit their findings to an online community. My students have gained information online through the World Wide Web as well as by e-mailing scientists, students, teachers, and pen pals throughout the world. When students from different countries exchange information about plant use in their respective cultures, a rich exchange can occur; I heartily encourage such encounters.

A Word about Safety

Not all plants are safe to work with. Some cause skin outbreaks or irritation (such as stinging nettle [*Urtica dioica*] and poison ivy [*Rhus radicans*]) and others are poisonous or even fatal if swallowed. Before handling a plant, you should make sure that it won't cause skin irritation or an allergic reaction. You should not eat any plant unless it is a common food substance that you can buy in a grocery store or supermarket. **It is not safe to eat wild plants** unless you have a trained expert to help you correctly identify them. Before working with any plant, a good background study should be made into the characteristics of that species. **Unsafe and illegal plants should be avoided**. Directions for background research can be found in Unit 2.

With the exception of Unit 7, on experimentation, none of the laboratory activities described in this book involve ingestion of plant substances. If you do want to create an experiment involving the ingestion of plants, you should read Unit 7 carefully along with the safety procedures. There are some stringent guidelines for experiments involving eating plants or drinking tea. In some cases it is permissible to eat a common food plant as part of a science experiment. For example, my students have done experiments to test whether mint-flavored gum helps freshen breath, whether tamarind juice can soothe a sore throat, whether raspberry leaf tea helps to ease menstrual cramps, and whether chamomile tea helps children fall asleep at bedtime, among other experiments that do involve drinking or eating a plant substance. In all of these instances, we made sure, as you also must, to **follow the safety procedures outlined in Unit 7**.

There are some plants that consistently produce exciting results during the course of conducting the activities in this book. They are safe and effective, and have a large number of uses. Some of these favorites are garlic, lemon, ginger, soybeans, cinnamon, aloe, and coconut; more plants like these can be found in Appendix B. However, since a large aspect of this book is geared toward studying a plant from the student's own culture, I always allow students to select their own plants, provided they are legal, safe, and easily obtainable.

Acknowledgments

I would like to thank many people for their friendship, assistance, and encouragement during the course of this project. They include my mother, Eleanor A. Magid, for always believing in me, and my other family members: Robert, Kirsten, Bjorn, Katya, and Maja DeBear. I thank my friends Jonathan Ball, Bill Ganter, Michael Squires, Ellen Resnick, Cheryl Bezis, Terezima Santos, Lisa Metropolis, Karen First, Nancy Nowak, Laura Sylvan, James Paye, Susan Wheelock, Katie Kennedy Cavanough, Pat Kranish, Lois Dodd, Hannah Orden, and Marie Arria. Thank you to Susan Frayman and Joy E. Runyon of The New York Botanical Garden Press, Elysa Hammond, and Michael Balick for their help in publishing this book. To my headmaster Donald Pellegrini and my department head Edmund Sprissler I am grateful for the support and freedom to be creative and try new things in the classroom. Thanks also to my colleagues at work, especially Jeannette Sisco, Krishna Rajangam, Cliff Whitehead, Andy Soo, Marian Sweeney, Bob Capuano, Mary Ellen Corbett, Grace Diggs, and Mary Dwyer. I am indebted to my first ethnobotany teacher, Gregory Anderson at the University of Connecticut (his was my favorite course), to Andrew Grellor, a botanist at Queens College, and to Bruce Benz, who taught me so much about ethnobotany during an Earthwatch expedition in Mexico. I appreciate the support of Karan Talentino and Kay Dunn at Simmons College and all of the EnviroNet participants. Many thanks also to my former students in Liberia, especially Cyrus Paye, Dopoe Menkarzon, John Barleah, Sekou Kamara, Solomon Sangaray, and Abraham Kamara; and to my students at West Roxbury High School for their outstanding work in plant science, especially Mia Valentini, Fidelito Gabriel, Melissa Rodriguez, Corey Ricker, Griselda Blanco, Nevada Saverse, Martine Amazon, Ralston Queensborough, Jennifer Phillips, Veronica Santa, Lisa Leitao, Asmaru Sinense, Raphaella Exat, Lissette Rivera, Jennifer Garcia, Luis Juarez, Lunhide Amazon, Jalela Austin, Argensis Cordonessosa, Shakir Amin, Rosa Jusino, Miguel Gonzalez, Christina Pruden, Maynor Sanchez, and Jerry St. Cloud.

A note concerning the illustrations: Except where indicated otherwise by a credit line, the photographs and drawings throughout this book were done by the author.

And finally, this book is dedicated to my daughter Amity and her friends Cleo, Chessie, Christelle, Eleanore, Alyse, Brendon, Isabel, Marianna, Noel, Dana, Margo, Silda, Nina, and Kalina; and to my own students and all young people. My hope is that this book will have at least a small impact in the struggle to alleviate the two gravest threats to our children's future — environmental degradation and war — by encouraging people to preserve the rich diversity of the plant world and by advocating multicultural tolerance and appreciation in our global community.

photograph by Charles M. Peters

Unit 1
Introduction to Ethnobotany

What Is Ethnobotany?

Ethnobotany is the study of the relationship between people and plants. The word *ethnobotany* is a combination of *ethno*, meaning "people" or "cultural group," and *botany*, meaning "the study of plants." An ethnobotanist may study how people collect wild foods for a meal, use herbs to treat illness, or craft canoes from plant materials in the forest. The opportunities for ethnobotanical study are almost limitless, be it in the laboratory or in the field. (Above, ethnobotanist Elysa Hammond, sitting, conducts fieldwork with local tree expert Manuel Chota at Jenaro Herrera in the Ucayali River basin in Peru.)

Ethnobiology is the study of the relationship between people and other living things. Although many examples exist of animals, fungi, or microbes that provide people with useful products (such as wool, cheese, and penicillin), we find that the diversity of useful goods derived from the plant kingdom is far greater than the diversity of goods from other living things. For that reason, the focus of this book is ethno*botany*. However, if you would like to learn about a useful organism other than a plant, a good place to start would be Unit 2, on researching and interviewing, or Unit 7, on designing experiments.

Why Are Plants Critical to Our Lives?

Plants Produce

Plants are primary producers. Through the process of photosynthesis, plants are able to transform water, gasses, and small quantities of minerals into living matter. All forms of animal life are consumers. Consumers, like ourselves, depend on plants, either directly or indirectly, as their source of food.

In addition to food, plants provide people with building materials, fibers for textiles, perfumes, ornamentation for house and landscape, gums, dyes, cosmetics, and beverages including the flavors for cola, root beer, and ginger ale. And because plants can't run away from their enemies like most animals can, they also produce a vast array of biochemicals as defense mechanisms against predators. Many of these chemicals have provided people with important medicines including aspirin, blood pressure medication, and birth control pills.

Plants produce the oxygen we need to breathe as a byproduct of photosynthesis. They take in carbon dioxide (the air we exhale) and, fueled by sunlight energy, combine that CO_2 with water to make simple sugars. Oxygen is released in this process and made available for use by people and other living creatures. During the process of respiration, animals and other living things inhale this oxygen and metabolize the sugars and other carbohydrates made by plants, thereby gaining the energy they need to work and play.

Plant communities such as forests also provide us with other vital environmental services. They supply food and shelter for wildlife, recycle water through transpiration, trap CO_2 (the gas that causes global warming), and protect the soil from washing away with the rain. Acting like large, natural air conditioners, plants cool us during the summer and modify temperature extremes during winter. They even intercept air pollutants and act as filters to cleanse water systems.

Healthy plant communities are vital to the well-being of our planet. Scientists have found that where large-scale deforestation takes place (as in the photograph below), flooding increases during the rainy season and droughts become more acute during the dry season. Without forest cover, soils become more compacted and rainwater runs off the surface rather than entering down into the soil profile where it can be stored for later use. When soil is eroded from exposed fields and washes away in rivers and streams, the land becomes less fertile, less productive. The eroded sediment that moves downstream also degrades aquatic ecosystems; in fact, in coastal areas where there has been extensive forest clearing, fish habitats and coral reef systems may be destroyed. Water systems near wetlands that have been deforested or developed often have higher levels of contaminants, when there are no longer enough plants to filter out pollutants and certain minerals that are harmful in higher concentrations.

photograph by Laura Sylvan

Our Dependence on Plant Diversity

Botanists have so far described about 250,000 species of plants worldwide; 50,000 more may await discovery. Any number of these species might provide us with new sources of medicine — perhaps even the cure for AIDS or cancer — but might easily become extinct before they are catalogued. Throughout the world, the pressures of development, urban sprawl, and the expansion of logging and corporate-based agriculture are steadily reducing global biodiversity, the rich variety of biological life. Currently one out of every eight plant species is vulnerable to extinction. Scientists estimate that about 1,000 plant and animal species now go extinct each year.

Why is plant diversity important? Whether you are considering genetic variation within a species or the variety of plants in a region, plant biodiversity may be nature's greatest gift to humankind. This is most evident in our food supply. About 12,000 of the 250,000 plant species documented have been used by people for food. Yet only 150 of those 12,000 plants have ever been cultivated widely, and a mere 20 species of plants provide 90% of the world's food! Clearly, a wealth of new food plants remain to be discovered and developed.

For all the major food crops, farmers have developed thousands of diverse varieties over the past several millennia by selecting and replanting seeds from the most desirable plants in their fields. For example, in India, at least 30,000 "land races," or folk varieties, of rice exist, each one adapted to a community's local growing conditions and selected for special characteristics. Scientists rely on the genetic diversity of these land races and the crop's wild relatives to develop new crop varieties.

A growing concern exists about the loss of these locally adapted varieties. As farmers replace their traditional varieties with modern, professionally bred seed purchased from large companies, local crop diversity is eroded. As large tracts of agricultural land become seas of genetic uniformity, we come to rely on a narrower genetic base for the majority of our food. A loss of genetic diversity leaves a crop more vulnerable to the spread of pests and disease. The history of the potato in Ireland demonstrates this in a dramatic way.

The potato originated in the Andean highlands of Latin American. There, more than 60 species of potatoes and thousands of varieties flourished in the cool mountain soils, a harsh environment where few other crops could survive. In the sixteenth century a few potatoes arrived in Europe — botanical curiosities that may have been taken by Sir Francis Drake from ships of the Spanish Armada. Eventually, the descendants of this handful of potatoes arrived in Ireland, a land where the cool, damp climate made grain production difficult and unpredictable. The potato plant thrived by comparison and soon became a staple of the Irish diet. By the mid-1800s potato cultivation in Ireland was extensive; people relied on it almost exclusively as their main source of food.

potato (*Solanum tuberosum*)

Originating from that first handful of potatoes, the potato fields of Ireland contained little genetic diversity and variety necessary to resist outbreaks of pests and disease. This narrow genetic base led directly to the famous Irish potato famine of 1846–1847. A fungal disease called blight arrived on European shores in 1845. This disease spread rampantly among potato fields in warm, damp misty weather, moving between fields perhaps one hundred times faster than other pests. One or two weeks after infection, all plants in a field would be left dead, the underground tubers collapsing a month later with wet rot.

So extensive was the loss of food those years that nearly one million people starved to death or died of starvation-related diseases. Another one and a half million people fled the country, initiating a pattern of emigration that continued for the next fifty years. Indeed, the lowly potato and its narrow genetic base in Ireland left a profound mark on world history.

If we are to feed a growing population, we must safeguard plant biodiversity as our storehouse of new types and varieties of food, medicines, and building materials. Genetic diversity is the basis for plant adaptation to environmental fluctuations, be it an outbreak of pests or a change in weather patterns. The challenge for the ethnobotanist is to demonstrate and discover the many uses of plants and to promote the sustainable development of these riches for present and future generations. Our survival is dependent on our good stewardship of the world's plant resources.

What Do You Get When You Cross A . . . ?

Genetic engineering is a relatively new field of science that relies on genetic diversity. Genetic engineering seeks to take desirable genes from one species or variety and insert them into another in order to apply those desirable traits. For instance, a group of scientists isolated the "antifreeze gene" in a fish that naturally lives in cold water, and they were able to insert that gene into a tomato to create a plant that won't freeze in cold weather.

Some scientists are concerned that there may be unexpected detrimental consequences of altering the natural genetic makeup of organisms. A study at Cornell University, for instance, found that a certain corn plant that was genetically engineered to produce its own "pesticide" also produces a toxin that is harmful to Monarch butterflies, a beneficial pollinator. These scientists argue that a "super" species could unintentionally be created that might escape and take over the habitat of many other species; more stringent regulations and research, they warn, is needed if we are to avoid undesirable consequences of this technology.

Whichever side scientists take in this debate, there is nearly unanimous agreement that it is important to preserve the biodiversity we have so that we will not lose useful species or harm the gene pool that exists.

Plants as the Basis for Civilization

Following the human trek from the Stone Age to modern society, plant resources have played key roles at every step of development. Describing the impact of plants on the rise of civilization, global exploration, and modern medicine would fill many volumes, so in the following section we offer a few of the highlights.

Origins of Agriculture

Milestone developments in human history have often followed changes in the use of plant resources. Beginning more than 30,000 years ago, early hunter-gatherers relied on wild plants and animals to survive, and, over the millennia, they accrued an extensive knowledge of the location and growing habits of useful flora. With the rise of agriculture about 10,000 years ago in the eastern Mediterranean and the concurrent development of agriculture in Northeast Africa, Mexico, Peru, and Southeast Asia, people's relationship with plants grew increasingly organized. People began to cultivate wheat, barley, lentils, and peas, and they soon developed special tools for harvesting and processing grains. As agriculture supplied people with a reliable source of storable food, they began to settle in villages and to eventually create towns some 6,000 years ago. Thus, people's interaction with plants forged the very foundations of civilization.

Plants and Global Exploration

Plant resources have played a tremendous role spurring global exploration. For the sake of spices, foods, and medicinal plants, explorers and merchants have sailed uncharted seas, colonies have been established, native peoples have been pushed off their land and enslaved, and wars have been waged.

The oldest botanical expedition on record is that organized by Queen Hatshepsat from Egypt to Syria in 1485 B.C., to collect frankincense (*Boswellia* species) and myrrh (*Commiphora* species) to use to embalm the dead. As a result of this trek, her husband and brother-in-law introduced many Syrian plants to Egypt. Certainly, this initial record of plant movement was merely a foretaste of future expeditions in search of plants that irrevocably changed the world.

In the Biblical and Classical periods, spices ranked among the most highly prized commodities, as precious as gold, silver, and jewels. Beginning in the thirteenth century, the search for spices played a major role in the great Portuguese, Dutch, and English voyages that, in essence, led to the expansion of European colonialismthroughout the entire world. Columbus sailed in search of a sea route to India and the Spice Islands and merely stumbled on North America en route.

cacao (*Theobroma cacao*)

In the Americas, a different set of plants fueled exploration, trade, and politics. In pre-Columbian times, cacao seeds (*Theobroma cacao*), introduced to Mexico from the Amazonian rain forest, were so highly valued that they were traded like coins of gold. They served as a currency of the Aztec empire for centuries before the arrival of Europeans; Aztec ruler Montezuma stored cacao seeds like treasure.

A Short History of a Long-Time Favorite (Chocolate!)

When Cortéz conquered the Aztecs, he returned to Spain with treasure sweeter than gold — cacao seeds and the recipe from Montezuma's court. Indeed, the scientific name of cocoa — *Theobroma cacao* — literally means "food of the gods." Served with honey and sugar, Aztec *chocolatl* so delighted the Spanish royalty that they kept it a secret for 100 years. When Spanish princess Maria Teresa gave her fiance, the French king Louis XIV, a gift of chocolate in 1615, the secret was out and chocolate made its way to the rest of Europe. The first chocolate factory was established in the United States in 1779. Annual consumption of chocolate in the United States is now 12 pounds per person.

Native plants became the seeds of world commerce. Rubber seeds (*Hevea* species) were stolen from Amazonia to create extensive plantations in Southeast Asia. African coffee (*Coffea* species) was introduced to Central America, eventually to become the mainstay of their economy. Valuable plant resources were hidden, sequestered away and grown elsewhere to enrich colonial powers. European botanical gardens served as repositories of these stolen treasures.

In 1630, Jesuit priests in Peru learned from local forest dwellers of a medicine called quinine found in the bark of the cinchona tree (*Cinchona* species) that grew on the eastern slopes of the Andes. Quinine served as the sole treatment of malaria, a deadly tropical disease transmitted by mosquitos. This cure for malaria allowed explorers and soldiers to survive in the tropics, enabling further colonization efforts and eventually affecting even the outcome of World War II in the Pacific theater. In 1944, scientists succeeded in synthesizing the chemical quinine, making a cure for malaria more widely available.

Currently, the quest for a treatment for AIDS and cancer continues to fuel the exploration of plants from remote, scientifically unexplored areas. The close relationship between plants and medicine is discussed in greater detail in the following section.

Plants and Medicine

The science of botany began as a branch of medicine. Indeed, nearly all formal studies of plants prior to the early Renaissance concerned themselves with plants that had medicinal properties. Collectively, these books are known as *herbals*. Herbals were collections of accumulated plant knowledge and lore, compiled from prior works and oral histories, and often augmented with the observations and experiences of the author. The chronology on pages 6–7 shows key historical developments in the evolution of ancient herbals and subsequent developments in the botanical sciences.

The Discovery of a Drug for AIDS from a Samoan Tree

Ethnobotanist Paul Cox works with traditional healers in Samoa studying their use of plants as medicine. One healer described the use of bark from the mamala tree (*Homoalanthus nutans*) to treat a form of acute hepatitis. In 1984, Dr. Cox sent a sample of this plant to the National Cancer Institute for analysis. In the lab, a team of scientists found that extracts from the wood stopped the HIV-1 virus from infecting healthy cells. Further work is being conducted on the plant at this time for the development of a drug that may eventually become effective as part of a combination therapy.

Works of medical-botanical origin have been produced throughout the world since the invention of writing. However, the depth and complexity of the earliest herbals clearly reflect a tradition that was established long before such information could be recorded. The earliest known herbal, the *Pen Tsao Ching*, composed by the Chinese emperor Shen Nung in 2700 B.C., contains 365 prescriptions including the use of such well-known medicinal plants as ginseng, the opium poppy, and ephedra (the source of ephedrine, a drug commonly used to treat asthma).

Among the most complete of the first written herbals is the Egyptian *Papyrus Ebers*. Although the earliest remaining copy dates from about 1550 B.C., it contains information that may have been written up to 2,000 years prior. *Papyrus Ebers* lists about 800 prescriptions including herbal remedies, antibiotics, and the preparation of cosmetics.

In the West, Greek scientists made significant contributions to the field of botany. The Greek physician Hippocrates (c. 460–377 B.C.), best known for the physician's oath that bears his name, compiled a list of useful plants and their value to medicine. Later, Aristotle (c. 384–322 B.C.) began systematic observations of plants and animals. His student Theophrastus (c. 371–287 B.C.), "the father of botany," inherited his library and continued his work, concentrating solely on plants. The ninth volume of his *Enquiry into Plants* contains a significant amount of information on medicinal plants.

However, one of the most important and influential of the herbals was *De Materia Medica* written by Pedanius Dioscorides, a Greek physician of the first century. This herbal, an illustrated compilation of more than 500 medicinal plants, became the most important reference on medicine and pharmacology for the next 1,500 years, until the Renaissance, and was also responsible for the eventual development of botany as a science in its own right.

Chronology of Some Milestones in Plant Science

Year(s) and region	Event
B.C.	
c. 8000, Middle East	Agriculture begins with the cultivation of wheat and barley.
c. 6000, Mexico	Agriculture begins in the Americas with the cultivation of maize, chili pepper, and squash.
c. 5000, China	Rice and cabbage are domesticated.
c. 4000, Pakistan	Cotton is cultivated.
c. 3000, Egypt	First paper, cosmetics, and antibiotics were developed. Medical text lists 800 drugs.
c. 2700, China	Emperor Shen Nung compiles the first herbal.
c. 2000, India	*Rig Veda*, a book of folk poems refers to medicinal properties of plants.
1485, Egypt	Queen Hatshepsat organizes first recorded botanical expedition, to East Africa to find myrrh and frankincense.
c. 500, Babylonia	Pollination is understood for the first time.
460–377, Greece	During his lifetime, the physician Hippocrates prepares a list of medicinal plants and builds a tradition of objectivity to help distinguish fact from superstition.
c. 300, Greece	Theophrastus writes a comprehensive botanical text, the *Historia Plantarum*.
c. 250, Belize	The Maya are cultivating cacao.
A.D.	
c. 77, Greece	Dioscorides writes *De Materia Medica*, describing and illustrating 500 medicinal plants.
335, Europe	The first appearance of cloves in the West is recorded, as the spice is brought to the Roman emperor Constantine.
570–632, Arabia	Most of the active spice trade routes throughout the Arab world, northern Africa, India, Malaysia, and Indonesia pass through the city of Mecca. Mohammed, the founder of the Islamic religion, lives in Mecca during this period, and this central location on a heavily traveled route is part of the reason for the subsequent spread of Islam.
610, Asia	Papermaking is introduced to Japan from China.
c. 1020, Arabia	The physician Avicenna writes a book listing 700 plant medicines.
c. 1190, Italy	A physician at the School of Salerno, Europe's leading medical center of the time, writes a book of "simples," primary ingredients from which prescriptions were formulated; the manuscript becomes the prototype for the modern herbal and pharmacopeia, and is the first document where an attempt was made to standardize plant names.
1214–1294, England/France	Roger Bacon is among the first to declare that knowledge comes not from authority and reason, but from firsthand experience; he uses mathematics, physics, and observation of nature to test ideas.
1315–1317, Europe	With crops at less than one-half their normal production in 1315, the worst famine of the Middle Ages occurs when people resort to eating seed intended for the following year's crops.
1489–1566, Germany	Brunfels, Bock, and Fuchs create botanical illustrations by observing living plants, rather than continuing the common practice of copying previous drawings that were, in turn, copies of previous drawings.
c. 1487–1880, Europe	Countries such as Portugal, Spain, France, Holland, and Britain sent expeditions to find spices, and colonized countries in large part to find and grow spices, sugar cane, and other plants.
1525, England	The earliest known English herbal is printed.
1533, Italy	Botany is established as a discipline separate from medicine when a professorship in botany is created at the university in Padua.
1595, France	Deforestation causes bakers in Montpellier to use bushes to fire their ovens; dwindling forest resources begin to create energy shortages throughout Europe, and an eventual shift to coal.
1610, Europe	The practice of drinking tea is introduced to Europe.
1630, Peru	Jesuit priests learn from native people about the cinchona tree's use as a cure for malaria.
1634–1637, Holland	The Dutch mania for tulips raises the price for a single bulb to as much as the equivalent of today's $50,000, until the market collapses in 1637, nearly ruining the entire Dutch economy.
1665, England	Robert Hooke writes about the structure of cork as viewed through a microscope — he is the first to describe a "cell."
1706, Sri Lanka	Coffee trees are shipped to the botanical garden in Amsterdam; only one survives, and it becomes the parent tree of the many thousands that were shipped to South America to create the first coffee plantations there in the late 1700s.

1737, Sweden	With the publication of *Critica Botanica*, Swedish botanist Carl Linné (known as Linnaeus) develops and sets forth rules for his binomial system for naming plants.
1768, England	Joseph Banks takes part in an expedition to South Sea Islands to explore plant and animal life.
1779, Holland	Jan Ingenhousz publishes *Experiments upon Vegetables*, in which he is the first to describe plant respiration and photosynthesis.
1785, England	William Withering publishes his discovery of the medicinal use of foxglove to cure certain heart ailments.
1789, France	Antoine Laurent de Jussieu publishes *Genera plantarum*, which combines Linnaeus's nomenclature and Adanson's natural classification system. This work becomes the basis for plant classification as we know it.
1804, United States	The Lewis and Clark Expedition begins. Lewis was prepared for the exploration by studying botany with Benjamin Smith Barton for nine months.
1826, Myanmar	The British conquer "Burma," largely for its vast wealth of unexploited forests. By the end of that century, 10 million acres of forest will have been cleared for teak.
1839, United States	Charles Goodyear invents vulcanization, a technique for processing rubber; what will follow is a rubber boom in Brazil, whose exports of the material will rise from 31 tons in 1827 to more than 27,000 tons by 1900.
1845, Ireland	Potato blight reaches Europe from the Americas. The Irish potato crop will completely fail the following year, and one-quarter of that country's population of 8 million will either die or emigrate.
1859, England	Charles Darwin publishes *On the Origin of Species by Means of Natural Selection*, in which he asserts that a given species can change gradually over many generations and result in an organism that is so distinct from the original as to be called a new species.
1866, Austria	Having experimented with the breeding of pea plants since 1857, Gregor Mendel publishes his discoveries about trait inheritance and the variation of traits among offspring of the same parents; largely ignored when it is first published, his work will later become the basis of genetics.
1869, United States	A biologist imports European gypsy moth to the United States for study; a few of those insects escape and will go on to establish populations that cause great devastation to eastern U.S. forests.
1883, Ethiopia	Addis Ababa is made the capital. Within twenty years, the city's demand for charcoal will cause all the trees within a 100-mile radius to be felled.
1895, United States	The term *ethnobotany* is coined by botanist John W. Harshberger.
1900, Micronesia	The British-owned Pacific Islands Company purchases mineral rights to Ocean Island for £50 a year; over the next eighty years, the mining of 20 million tons of phosphate for agricultural fertilizer will decimate the native vegetation and destroy the homeland of the island's 2,000 residents.
1912, United States	Frederick Hopkins discovers that food contains chemical compounds (in addition to fats, carbohydrates, and minerals) that are essential to human health; these will later be known as *vitamins*.
1921, United States	George Washington Carver gains national fame when he demonstrates some of the many uses for peanuts during testimony before the Congressional Ways and Means Committee to advocate a protective tariff on the importation of this legume.
1931–1939, United States	Excessive clearing of grasslands, subsequent soil cultivation, drought, and overgrazing of cattle combine to create the Dust Bowl — an area covering parts of Colorado, Kansas, Oklahoma, Texas, and New Mexico where heavy winds eventually remove the top five inches of soil from 10 million acres of land and 3.5 million people abandon their farms.
1937, Hungary	Albert Szent-Györgyi is awarded the Nobel Prize for his discovery of vitamin C, which he isolated while studying paprika.
1937–1991, South America	Richard Evans Schultes does fieldwork in the rain forests of the region to learn about medicinal and otherwise useful plants from local people; he is widely regarded as the founder of modern methods in ethnobotany.
1952–1957, England	Crick, Watson, Wilkins, and Franklin make a model of the structure of the DNA molecule and also discover its chemical structure.
1957, United States	In clinical trials, extracts from the common periwinkle are found effective in the treatment of childhood leukemia.
1962, United States	Rachel Carson publishes *Silent Spring*, in which she raises serious concerns about the effects of pesticides on wildlife and the environment; the book spurs a vigorous new era of environmental awareness.
1972, United States	Citing toxicity concerns, the Environmental Protection Agency bans most uses of the pesticide DDT.
1994, United States	The Food and Drug Administration approves the first genetically modified food — Flavr-Savr tomatoes, designed to ripen more slowly and have a longer shelf life.
2000, Canada	An international agreement is reached that will require shipments of genetically modified crops to carry a label stating that they "may contain" genetically modified organisms and are not intended for introduction into the environment.

With the invention of printing, and the fleeing of Greek scholars westward following the fall of Constantinople, *De Materia Medica* made its way to western Europe. Renaissance botanists there found that the plants of their regions differed from those described in *De Materia Medica*. This discrepancy produced confusion until scholars realized that different regions produced different species. With that now-obvious insight, the science of botany began to blossom and take off on its own from there.

The development of a system of scientific names in Latin, the discovery of photosynthesis, the identification of the cell, the theory of evolution, and the understanding of plant genetics all brought major new insights to the field. However, it is important to remember that botany originated as a branch of medicine. It is that fundamental bond between botany and medicine that this book harkens back to in the experimental units presented later.

Modern-Day Ethnobotany

Ethnobotany continues to thrive in modern history. There are many approaches modern ethnobotanists may take in their study of plants. One of the most intriguing and promising areas of research is in the study of medicinal plants. One out of four prescription medicines in the United States contains active ingredients originally derived from plants.

Many of these medicines were discovered using an ethnobotanical approach. That is, scientists learned of the active plant compound by tapping into the accumulated wisdom of "folk medicine" — an understanding of useful plants (such as that of the herbalist doctor in the picture at left). When ethnobotanists learn of a plant medicine that seems effective, they can use lab techniques such as *bioassays* or *chemical analyses* to identify the bioactive plant components that may have medical applications. Applying the scientific method to the traditional knowledge of indigenous cultures is one way of testing the merits of specific practices in folk medicine.

While many ethnobotanists study medicinal plants, some look for plants that provide other useful products such as fibers, foods, beverages, and perfumes. Seed companies, for example, might consult an ethnobotanist to assist in the search for new crop varieties that are more appealing, tastier, or more disease resistant than others currently on the market.

Some ethnobotanists study plant origins. Using carbon dating and other techniques, they discover what plants were used by prehistoric peoples and how agriculture may have developed in different regions of the world.

Given their knowledge of plant diversity and their appreciation for the accumulated wisdom of traditional cultures, most modern ethnobotanists have a strong sense of responsibility to protect both people and plants throughout the world. Given the accelerating rate of deforestation, most traditional cultures that still actively employ plants as medicine are under threat. Ethnobotanists recognize that efforts to conserve biodiversity must also include cultural diversity, and thus they often involve themselves in protecting the rights of indigenous people.

In the following section, we present portraits of three modern-day ethnobotanists: George Washington Carver, Delfina Cuero, and Richard Evans Schultes. Each of these three people has made significant and unique contributions to our current understanding of useful plants.

George Washington Carver: Using Ethnobotany to Assist Rural Farmers

George Washington Carver is best known as a scientist, inventor, and agricultural chemist who developed over 300 uses for the peanut. Although rarely described as such, Dr. Carver was also an ethnobotanist. Working closely with local farmers, his approach to research in the field often relied on ethnobotanical methods. He studied how people raised crops and the diverse ways they used plants in their daily lives. And, like all good ethnobotanists, Dr. Carver's primary concern was to find ways of using plants that would help people.

Dr. Carver was born of slave parents in Diamond Grove, Missouri, near the end of the Civil War. As an infant, George and his mother were kidnaped by Confederate raiders and taken to Arkansas. Rescued by a neighbor and returned to the Missouri farm, infant George was raised by Moses and Susan Carver as their own son. During his childhood on this rural estate, George developed a deep love for nature and a fascination with plants that remained with him the rest of his life.

George Washington Carver worked diligently to get an education at a time when African-Americans were still excluded from most educational institutions. He received his Bachelor of Science degree from the Iowa Agricultural College in 1894 and a Master of Science in 1896. He went on to serve as an instructor at the Tuskegee Institute in Alabama, where he remained on the faculty until his death in 1943. His work on the industrial uses of plants led to the development of more than 118 different products, including a rubber substitute, and more than 500 dyes and pigments from 28 different plants. Three separate patents were issued for his invention of paint and stain made from soybeans.

His understanding of local farming practices and the acute problems of soil degradation in the rural South served as a guide for his work. Farmers desperately needed new crops to replace the revenue received from the soil-depleting crop cotton, which suffered severe damage by the boll weevil in the early 1900s. To restore soil fertility, he promoted the use of legumes (peas, beans, and peanuts) in crop rotation, to add nitrogen to exhausted soils. Likewise, the use of sweet potatoes in crop rotation and the planting of pecan trees helped restore soil health.

Due to Carver's influence introducing crop rotation, large surpluses of peanuts (*Arachis hypogaea*) and sweet potatoes (*Ipomoea batatas*) soon began to accumulate. To develop new markets for these crops, he began the work he is most well-known for — developing 325 uses for the peanut. Additionally, he discovered 105 uses for sweet potatoes and 75 applications for the pecan (*Carya illinoinensis*).

Going Nuts

George Washington Carver found that the peanut — a nutritious, high-protein, oil-rich seed — could be developed into a myriad of industrial and edible products including the following:

adhesives	axle grease	bleach	buttermilk
charcoal briquettes	cheese	chili sauce	creosote
dyes	flour	ink	instant coffee
laundry soap	linoleum	mayonnaise	meat tenderizer
metal polish	milk flakes	paper	salve
soil conditioner	shampoo	shoe polish	shaving cream
synthetic rubber	talcum powder	wood stain	worcestershire sauce

His work grew out of an ethnobotanical approach to studying plants. One of his favorite pastimes was to travel about the Alabama countryside talking to farmers, learning as much as possible about their planting techniques and uses of plants. He learned about the peanut, which enslaved Africans had brought with them to America, from the people in his community; in return, helped make this humble, nutritious plant one of the world's most important crops.

Delfina Cuero: An Ethnobotanist of the American Southwest

Delfina Cuero was a Kumeyaay Native American woman whose deep and complex knowledge of Southwestern plant use inspired anthropologist Florencia Connolley Shipek to put her fascinating but otherwise anonymous story in print.

Delfina Cuero, born about 1900, was one of the Kumeyaay people who originally inhabited the area near San Diego, California, and the area just south of the United States border with Mexico. Forced to leave their homeland in 1910, some of the Kumeyaay were driven onto reservations while others were left to live as migrants. Delfina's family was one of many who lived as wanderers, moving wherever there were wild foods to gather or work to be done.

While the men engaged in agricultural labor, women foraged for wild berries and foods to feed their families. The ranchers her father worked for would pay them with old clothes or food, though never enough food, rather than money. Their houses — small, hand-made huts — were fabricated from native willows, reeds, and yucca fibers. Delfina, early on, understood that wild plants were essential to her family's survival.

Driven further out toward the mountains and into Mexico, Delfina became adept at locating a host of edible and medicinal plants that grew around her. She learned to use cacti, acorns, pine nuts, seaweed, and wild greens as food. She pounded flour from wild cherries, mesquite berries, lilac seeds, and grass. Manzanita trees provided berries for drink and wood from which to carve ladles and tools. Hunting and fishing rounded out their diet. Traps woven from agave fibers enabled the family to catch fish, and the same agave fibers provided fabric for homemade sandals. Fruits, vegetables, and meat were dried and saved for winter.

As an adult, Ms. Cuero learned to use herbs to treat the common ailments of her children, relatives, and community. To cite a few among many, she used spice bush for toothaches, yarrow for skin problems, sweet fennel for stomach disorders, and black sage for bodyaches. Widowed early, and with five children, her knowledge of plants grew ever more important for the health and survival of her family.

Over time, the increasing number of people living in the mountains made it harder to live off the land. Ms. Cuero began to support herself with supplemental income from washing and ironing clothes; she refused to become dependent on federal assistance for survival. Eventually she made her way to Mexico in search of food and work. There, in 1968, she met anthropologist Florencia Shipek, who, impressed with her ethnobotanical knowledge, published her story, based on a series of interviews, in a book called *Delfina Cuero: Her Autobiography.*

In her later years, Ms. Cuero moved to a reservation house for the elderly. She continued daily walks in the mountains to collect wild foods and medicinal plants. She propagated plants in her garden and guided botanists and anthropologists on walks to teach them the lore and uses of native Southwestern plants. Despite her difficult life, she was always a cheerful, generous person and she used her complex botanical knowledge to enrich the lives of those around her.

Richard Evans Schultes: Establishing the Foundations for Academic Ethnobotany

Richard Evans Schultes, considered the father of modern ethnobotany, has served as a model and inspiration for this profession throughout the world. Beginning with his undergraduate thesis fieldwork in Mexico in 1937, he worked extensively with native peoples in Latin America, studying their knowledge and use of plants. Traveling alone with little gear, he immersed himself in the lives of the people he studied. He pioneered the participant-observer method of research — he learned by sharing in daily village activities and participating in indigenous rituals involving plants. This approach allowed him to develop deep and meaningful relationships with the people he studied. His writings communicate his profound respect for these cultures and for their knowledge and use of natural resources.

During his fourteen-year residence in the Amazon, Dr. Schultes collected more than 25,000 plant specimens, an essential part of ethnobotanical research. He identified hundreds of medicinal plants, and described species used as fish poisons, arrow poisons, and hallucinogens by native peoples.

During World War II the U.S. government assigned him the task of studying the availability and abundance of native rubber trees in the forest. He was to assess the potential supply of native rubber for use by the Allied Forces. During this time, he also conducted studies of medicinal plant use that would later form that basis of his book *The Healing Forest.*

As a teacher of ethnobotany at Harvard University, Dr. Schultes inspired many students to carry on the practice of meaningful fieldwork with indigenous peoples. He also promoted an interdisciplinary approach to ethnobotany, working closely with chemists and pharmacologists to identify the bioactive compounds involved in native plant use. His influence on his graduate students, now professionals themselves, continues to resonate throughout the academic field of modern ethnobotany.

That's from Sap?!

What do airplane tires, chewing gum, and maple syrup have in common? All these products, and thousands more, are derived in part from plant sap.

Prior to World War II, all tires were made from natural rubber, derived from the sap of a native Amazonian tree, *Hevea brasiliensis*. The collected sap, actually a milky white latex, is made into rubber through a process called *vulcanization*, developed in 1839 by Charles Goodyear. Natural rubber has unique attributes of particular use to the transportation industry: It is strong and adherent, generates relatively little heat when flexed continuously (think of large trucks on bumpy highways), and remains flexible when exposed to subzero temperatures (such as those at high altitudes). Although most tires today contain synthetic polymers, natural rubber is still a key ingredient in their structure. For certain aircraft where tires are under tremendous pressure during takeoff and landing, such as the Concorde and the space shuttle, tires are made of 100% natural rubber.

Chewing gum comes from the sap of the chicle tree, *Manilkara zapota*, native to Central America. The Mayan Indians chewed chicle sap long before its milky latex was processed and sweetened for the first commercial sales of chewing gum in the nineteenth century. Today, chicleros still tap trees that grow in Central American rain forests.

In North America, some native peoples had long used the sugar maple tree as a source of sweetener. After slashing the bark of the trees in spring, that collected sap was boiled down to make a thick syrup. It takes approximately forty gallons of sap to make one gallon of syrup.

The Approach of This Book

The activities of this book are designed to teach you some of the ways that scientists investigate plants with the potential to discover new and helpful uses for humanity. This book also encourages you to conduct experiments on your own and assumes that you, too, may make new discoveries about a plant. Once you've completed the labs and activities found at the end of each unit, you can email the author to find out how to share your results with others: gdpaye@hotmail.com.

In this chapter we learned about ethnobotanists with different approaches to studying plants. Richard Evans Schultes traveled to remote Amazonian jungles to live with indigenous cultures that were not his own to study their use of plants. In contrast, George Washington Carver stayed right in his own backyard in Tuskegee, Alabama, and made discoveries about traditionally used plants from his own culture. Both men made great contributions to science.

Since few students have the opportunity to travel to distant countries to study plant use, the book takes the approach of George Washington Carver. I encourage you to study plant use in your own backyard, from your own family, neighbors, or culture. Study plants that are accessible. Study a traditional plant used by your grandmother, great-uncle, or a family friend. And if you are lucky enough to travel, this book may inspire you to explore the world as an ethnobotanist, learning about how people use plants wherever you trek.

George Washington Carver discovered 325 uses for the peanut, but only after recognizing that the peanut was a unique plant with great potential. He had done his homework and learned that this oil-rich crop could enrich the soil and nourish people. Likewise, your selection of a plant to study will make a big difference in the outcome of your experiments. If you do your background research carefully (Unit 2) before you choose a plant for your studies, you will have a better chance of getting exciting results.

Unit I Questions for Thought

On a separate piece of paper answer the following questions as completely as possible.

1. Why are more useful products made from plant materials than from animals?

2. How does the process of photosynthesis help our lives? What are the useful products of photosynthesis?

3. How do plants help to protect the environment?

4. Imagine that you are a biologist who has to address a group of U.S. senators about why we should protect a natural area. Write a one-page essay that explains why biodiversity is important to our lives and why the natural area should be preserved.

5. What sometimes happens when we base our source of food on a narrow range of species or varieties within a species?

6. What do you think an average day was like for people who lived in the days before there was agriculture?

7. Describe how the demand for the following useful plants and spices led to global exploration and trade:
   - frankincense and myrrh
   - cacao
   - rubber
   - coffee
   - cinchona

8. Why were the sciences of botany and medicine often studied together as one subject until roughly the end of the nineteenth century?

9. Many scientific developments are based on previous developments. For example, genetic engineering was not possible before the discovery of DNA and Mendel's laws of genetics. Using the time line in this unit, find another example of an invention or discovery that made subsequent inventions and developments possible. How are the earlier and later developments connected?

10. Which of the developments on the timeline do you believe had the biggest impact on modern society? Explain your answer.

11. How can common people's knowledge of medicinal plants help the scientific medical establishment?

12. Contrast and compare the techniques of the following botanists:
   - George Washington Carver
   - Richard Evans Schultes
   - Delfina Cuero

Unit I Activities

Field Exploration: Supermarket Botany

Name: ______________________________ Date: ____________________

Imagine that you are plant hunter embarking on a safari that takes you to the far reaches of your neighborhood supermarket. In the aisles of your grocery store, see how many plants and plant products you can find to answer the following questions:

1. Name as many products as you can that contain the following:

a. peanuts ____________________

b. coconut ____________________

c. corn ____________________

d. wheat ____________________

2. Is there an ethnic section (e.g., Hispanic, Chinese, Italian, or Jewish) of the supermarket? If so, list the cultures, the main foods and the plants they are made from below:

Culture(s)	Foods found	Plants used
____________	____________	____________
____________	____________	____________
____________	____________	____________
____________	____________	____________
____________	____________	____________

3. List foods that you find in the store from each plant part listed below:

a. roots or underground stems ____________________

b. leaves ____________________

c. seeds or nuts ____________________

d. flowers or flower buds ____________________

e. fruits ____________________

f. aboveground stems ____________________

4. Find a product from each of the following categories. Name the product and a plant that is listed as one of the ingredients:

Product	Plant(s) listed
a. hand/body lotion ____________	contains ____________
b. shampoo ____________	contains ____________
c. toothpaste ____________	contains ____________
d. cleanser or soap ____________	contains ____________
e. cereal ____________	contains ____________
f. nonprescription medicine ____________	contains ____________
g. beverage ____________	contains ____________
h. condiment ____________	contains ____________

Laboratory Activity: Supermarket Gardening

1. **Seed Collection.** Gather 5–10 types of raw seeds from the store or home for germination tests and planting. These may include various kinds of dried beans and peas, lentils, peanuts, sunflower seeds, bird seed, and seeds in fresh fruits like oranges, grapefruit, apples, avocado, bell pepper, winter squash, or pumpkins. You might also include seeds from dried spices such as coriander, fennel, or mustard seed.

2. **Root and Rhizome Collection.** Assemble a group of fresh edible roots or underground stems from the supermarket such as potatoes, sweet potatoes, yams, garlic, ginger, and cassava root. Sometimes you can get cuttings from these fleshy plant parts to take root in a pot of soil. You may also want to try rooting stem cuttings of watercress in a cup of water.

3. **Germination Tests.** Conduct germination tests for two or more types of seeds. Place at least 20 seeds of the same type on a damp paper towel on a plate and keep them covered and moist for a few to several days until seedling growth is visible. Only the viable (living) seeds will germinate. Estimate the percent viability of your seeds using this equation: number of seeds germinated ÷ total number of seeds × 100 = ____% germination.

4. **Planting Seeds.** Gather pots, potting soil, water-resistant ink, and labels. For each type of plant that you grow, create a legible label that states the planting date, the species or variety planted, and your initials. After putting the soil into the pots, add your labels and seeds or cuttings to each one. Water the soil. Sometimes it is helpful to cover the seeds and cuttings with plastic to prevent them from drying out until the new plants develop a strong root system. See how long you can keep your plants alive and how healthy you can keep them. Create a chart in your science notebook to allow you to keep track of what you planted, the dates that you started growing the plants, the plants' germination rates, and how well the plants grew, along with any other observations you think are relevant.

Unit 2

Choosing a Plant to Study: Interviews and Background Research

Introduction

This chapter will help you conduct the background research necessary to choose an appropriate plant to study. First, you will learn to conduct interviews about plant use with a knowledgeable relative, friend, or neighbor. Second, you will follow up your work with research in the library and on the Internet. Third, you will present your findings in a research paper.

Because this preliminary work sets the stage for subsequent laboratory experiments, the selection of a promising study subject is essential. Many of my students have chosen plants that produced exciting research results. Some plants exhibited antibacterial or antifungal properties, while other plants produced useful fibers for making papers. You might discover that your plant is very nutritious or that it yields a dye of a pleasing color. If you choose a plant about which little is known, you might make discoveries that no one before you has ever described. Or, if you choose a popular plant to study, you might be able to verify the value of traditional plant lore through your scientific experiments. Your research might even uncover new reasons to protect a threatened species or endangered cultivar and put you in contact with other scientists.

After reading this chapter and conducting interviews, if you are still having difficulty finding a plant to study, you may consult Appendix B in this book for suggestions. It lists some useful and interesting plants from ten different parts of the world. I chose these plants because they are relatively safe to experiment with, are widely used and/or easy to find, and are not endangered. However, they represent only a fraction of our planet's vast wealth of useful plants.

The Ethnobotanical Interview: Learning about Plants from Our Families, Friends, and Neighbors

Elders and Traditional Knowledge

Some people believe that anyone lacking an advanced college degree or specialized training has little to offer the scientific community. However, many professionals involved in ethnobotany today — botanists, anthropologists, ecologists, and medical doctors — deliberately seek out knowledgeable elders and healers from diverse cultures and traditions, to learn from them. With the scientific community now beginning to appreciate this wealth of practical information as a foundation for research, scientists are using scientific processes to test the validity of traditional knowledge. Frequently the wisdom of the elders is confirmed through such experimentation.

What kind of traditional knowledge might one expect when asking an elder about plants? They often know which wild plants are edible, which ones are poisonous, and which are medicinal. They may know how to prepare tonics, tinctures, and poultices for healing, and how and when to apply them. Your grandmother may know which tea is best to ease an upset stomach and which one is a remedy for poison oak. Someone else may know which plant makes a good soap and which ones work as insect repellents. Some people still weave mats or baskets from plants or remember how to build fishing traps. Each person's knowledge of plants will vary with their background, interests, and experience.

To many ethnobotanists, the loss of the indigenous knowledge of plant use is as critical a problem as the loss of plant diversity itself. This decline is not unique to any one particular culture or country. Under the pressures of modernization, the erosion of traditional knowledge is occurring throughout the world at an increasingly rapid rate. Elders themselves often feel sad that the younger generation seems uninterested in their heritage, and look for someone who can preserve the old knowledge before it dies out forever. Perhaps you will be one of the people who help save the knowledge.

Finding a Person to Interview

In this activity you are the scientist who is learning about plant use by interviewing an *informant* — someone with firsthand experience who can tell you what he or she knows. To find an appropriate person to interview in depth, you may need to talk with several different people. First, seek out older persons who are experienced gardeners, farmers, cooks, or healers. They may be someone from your own family and/or cultural background, or they may be someone from a different culture than your own. You could also interview a knowledgeable neighbor, teacher, or friend.

A Word on Keeping Safe

If you are interviewing someone you don't know very well or have just met, take a friend or an adult relative along on the interview instead of conducting the interview alone.

Don't go inside the house of someone you don't know — hold the interview outdoors if possible.

If anything the informant tells you during the interview makes you feel uncomfortable, trust your intuition and leave. There are plenty of other people to talk to about plants. Talking with an informant should be a comfortable and pleasant experience.

You might find an opportunity to raise the question of useful plants in a group setting such as a family reunion, a gathering of older relatives, or a family party. In such a gathering, raising questions about people's childhood use of a plant inspires reflection, reminiscence, and conversation. "Uncle Edward, what's in the juice you mix up for your garden to get rid of insects?" "Aunt Rosasita, what was that tea you used to give us to help us fall asleep?" "Grandma, what did your mother do for you when you had the flu?" Don't overlook the possibility of telephone interviews to talk to more distant relatives about plants too.

Another gathering place where the conversation is sure to revolve around plants is in a community garden. Community gardens grace the vacant lots of many large cities in the United States, and are rich sources of people to interview. Although the gardeners often represent diverse traditions, they are united by their common interest in plants. You might be able to tell the gardener's family background by the plants they cultivate: A Puerto Rican garden will surely boast hot peppers and cilantro; while in a West Indian garden you might find the vegetable kalalu and pigeon peas. Asian gardens may be adorned with climbing vines of bitter melons or winged bean, while an Italian garden might include a bright display of dandelion greens, eggplants, and basil. People often enjoy talking about the plants they raise and the different ways they use them (as these Greek and Irish community gardeners are shown doing here with the author).

photograph by Eleanor A. Magid

Conducting the Interview

When you have found someone to interview, always begin by introducing yourself and clearly explaining your objectives. You might tell your informant that you are studying ethnobotany and that one of the goals of your school project is to study a useful plant that is part of a family or cultural tradition. Further explain that you are to choose a specific plant for your research based on an interview with an adult who is well-acquainted with plants.

During the interview, be respectful of your informant's time and willingness to share information. A general guideline is to limit interview time to a half-hour or less. Don't press too hard for the person's age or any information they seem uncomfortable about giving you. If they hesitate to divulge something about plants, do not pry. Remember that in some cultures, knowledge of herbal remedies is kept secret. Always treat people's beliefs with respect even if you believe differently. If you are patient, sincere, and polite, the person is more likely to open up.

Use the questionnaire at the end of this chapter to guide your interview, moving through four basic steps (fours sections of the questionnaire) to collect information:

1. Gather background information about your informant (Section A).
2. Generate a general list of interesting plants (Section B).
3. Narrow the list down to the most promising species (Section C).
4. Ask more detailed questions on those select plants (Section D).

Don't worry if your interview does not follow this exact format; you need to be flexible. For example, if your informant launches into a childhood story about a healing plant, take notes and listen; you can get back to the other questions later. Keep in mind that the main goal of the interview is to identify a few promising plants for research. These steps are explained more completely below.

First, try to learn a little about your informant's background. Express interest in learning about their birthplace, native language, and most importantly, how they grew to know so much about and appreciate plants. Who did they learn from? Did they grow up on a farm?

Second, record all the plants mentioned by your informant in a preliminary list. When you ask about useful plants they're familiar with, they may reel off a long list of possibilities. On the other hand, they may know of only one or two. Write down as many as you can in Section B on the form.

Third, your goal is to identify a few (1 to 3) of the most promising plants to learn about in greater detail. To narrow down a long list of plants you may need to ask these types of questions: "Which of these plants are the most nutritious? Which of these plants make the most effective medicines? Do you have a favorite plant from the list you gave me?" Your goal is identify the most promising plants to focus on in last part of the interview.

Finally, you will ask more specific questions about these select plants to learn about their identification, use, and availability. If the plant does not grow in the region and you want to study it, you'll need to know whether you can buy it or grow it yourself. Include any stories or recipes they share with you. Write everything down or tape record the interview so you won't forget the important points.

If possible, collect detailed information about more than one plant, filling out a separate form for each one. The first plant may be unavailable or uninteresting to you once you begin the follow-up research. The more plants you learn about, the more you'll have to choose from when writing your research paper. To complete your interview, you may need to make more than one visit.

Before you go, be sure to thank the person interviewed for his or her time and help. Let the person know that you may need to return if you have further questions. You may also want to come back at a later date to share your research results with the person you interviewed or to take his or her picture to include in a project presentation.

What's in a Name?

Many plants occur in different regions of the world and are known by more than one culture. Given the differences in culture and language, one plant may have many different names. Before a systematic method was developed for giving universally recognized scientific names to plants and animals, you can imagine the confusion when, for instance, the same plant was known by one culture as "manzana," by another as "apfel," by another as "pomme," and by yet another as "apple."

Thanks to the work of the Swedish botanist Carl Linné (best known by his scholarly name, Linnaeus), every plant is given a Latin name. In his book *Species Plantarum*, published in 1753, Linnaeus established his system of *binomial nomenclature* —that is, the practice of giving each plant a two-part Latin name.

The first part of the binomial identifies the plant's *genus* (plural: *genera*), the group it belongs to on the basis of certain shared characteristics. The second part of the binomial identifies the plant's *species* (plural: *species*), the group of plants it belongs to on the basis of certain shared characteristics and with which it can be crossed to produce fertile offspring. In the case of the apple, you can tell from its scientific name, *Malus pumila* (genus *Malus*, species name *pumila*), that it is in the same genus as crab apple trees (such as Japanese crab apple, *Malus floribunda*, or Arnold's crab apple, *Malus arnoldiana*) but that it could not be crossed with a crab apple tree to produce fertile offspring.

Sometimes there is variation within the species. Variations created by humans, often for agricultural purposes, are called *cultivars*; variations that occur in nature are called *varieties*. For instance, Macintosh, Granny Smith, and Golden Delicious apples are all cultivars of the species *Malus pumila*. Although they look different, varieties and cultivars within a species can be cross-pollinated to produce new plants of the same species.

Background Research

After the interview, you will need to conduct background research on your selected plant. Using the library and Internet, you can find out if what you learned from your informant is verified in scientific or popular literature. A solid literature review on your topic will also assist you in making intelligent hypotheses about experiments that you will conduct later.

This section of the book will guide you in gathering background information about your plant and summarizing what you have learned in a research paper. Due to the importance of this research to your overall project, some brief guidelines on using the library and Internet, on taking research notes, and on writing the paper are presented below.

Find out More about Your Plant at the Library

Your school or community library should be your first stop in your search for information on your plant. You can browse the card catalogs, search computerized databases, and explore the natural science bookshelves; don't forget to look also for gardening books, herbals, and field guides. The librarians can help you expand your search to include magazines, government publications, newspapers, scientific periodicals, encyclopedias, specialized indices, and CD-ROMS.

Keywords that may help you in computerized searches include "ethnobotany," "economic botany," "medicinal plants," "herbs," "herbals," and the common and scientific names of your plant. Most libraries offer interlibrary loan services and, if they don't have a book you need, can request books on your behalf from other institutions. Specialized libraries like those found at many universities or botanical gardens have more detailed collections related to plants. If you intend to conduct really serious research on your plant, you might call and request permission to use such a collection.

Find out More about Your Plant on the Internet

If you have access to the Internet at school or at home, you will find a seemingly endless network of plant-related information sources. For example, a search for "medicinal plants" may turn up 50,000 related web sites. To begin your Internet search, you will need to find an online search engine, a program that will find web pages related to a keyword or phrase such as "Chinese food" or "aloe vera" or "plant ecology." (Some popular search engines include Altavista, Lycos, Yahoo, LookSmart, and InfoSeek, to name just a few.) The search engine will compare your keywords (called a *search string*) with its large database of web sites and locate the sites containing your search string.

The overwhelming amount of information accessible on the Internet makes it difficult to discern between reliable sources and those that may be misleading or dishonest. For example, a company selling garlic pills might claim that their product can cure everything from hang nails and colds to arthritis, diabetes, and cancer. How do you protect yourself from such fraudulent claims? First, do not rely solely on the Internet for your report references. Second, limit your online searches to reputable research institutions such as museums, universities, botanical gardens, and trusted environmental organizations. These sites may also provide a list of links to sites that have more in-depth information.

Finally, you might try contacting a scientist who specializes in your area of interest via e-mail addresses you come across on the Internet. Likewise, you can also contact other students via web sites that promote student–student and student–teacher collaboration, such as EnviroNet (sponsored by Simmons College), the National Science Foundation, and Access Excellence (sponsored by Genentech).

How to Take Notes about Your Plant

By now, most students have written some type of research paper for a class at school. The standard method of taking notes also works well in ethnobotany — that is, putting each topic and its reference on a 3" × 5" index card or half sheet of paper. This method of gathering information accomplishes two important goals. First, you can more easily organize your notes and ideas when you write your paper. Just sort the index cards to fit your outline or table of contents. Second, the index cards show your teacher that you did the research and wrote the paper on your own. Your notes help protect you from charges of plagiarism, accusations that you copied your final report from another book or article.

Each index card should contain only one topic from a single reference. At the bottom of the card, cite your source including author, title, date, publisher, place of publication, volume (if available), and page number(s). If it is an article in a journal or magazine, include the journal title, volume number, month, and issue number. If you use the same article or book for more than one note, you need only to cite the author, title, and page number on subsequent cards from the same source. Write on only one side of the paper so you can more easily sort your cards later.

The notes you take, and the final paper you write, should be paraphrased in your own words, to demonstrate that you understand the content of what you have read. Direct quotes should be identified with quotation marks. Here are two examples of index cards by students at West Roxbury High School in Massachusetts.

The History of Vanilla.
Vanilla comes from Mexico and is a beautiful yellow orchid. The Aztec Indians used it for flavoring. It was brought to Europe by the Spanish adventurer Hernando Cortés.
Mulherin, Jennifer. Spices & Natural Flavouring. Tiger Books International, London. 1994, p.94.

Medicinal properties of Green Tea
(Camellia sinensis)
It can reduce tooth decay because it has a lot of fluoride. A tea poultice can stop itching and swelling of insect bites. Far East Scientists say it reduces chances of getting stomach cancer and Green Tea boosts the immune system.
Ody Penelope. The Complete Medicinal Herbal Dorling Kindersley, New York. 1993, p.44

Writing a Research Paper on Your Plant

It is time to sum up your interview and research notes in a formal report. This report encompasses three main objectives: (a) to summarize what you have learned about your plant from the library and Internet; (b) to discuss how your interview results confirm or conflict with information in the literature; and (c) to help you develop hypotheses about laboratory experiments you will conduct later in this project.

Select a topic for your ethnobotany/plant report that is manageable in size. For example, the topic "Plant Life of Puerto Rico" is too broad. A more specialized topic such as "Cilantro (*Coriandrum sativum*) and Its Use in Puerto Rican Culture" would be more appropriate.

Your teacher can provide you with guidelines for developing a well-written paper in which one idea leads logically to another. Remember, it always takes several drafts or revisions to make a paper read clearly and smoothly, and it's always a good idea to read your paper aloud to someone or have someone read it aloud to you; problems with logic, flow, and grammar are easier to catch this way. Like all reports, your paper should include the following, in this order:

1. Title page
2. Table of contents
3. Introduction
4. Body of the paper
5. Summary
6. Bibliography

You may also add drawings, photographs, and graphs to your paper. To find out how to draw and photograph your plant, see Unit 3; for information about collecting, analyzing, and graphing data, see Unit 7.

Prepare a detailed outline to help you organize your notes before writing. An ethnobotanical report should include, if possible, the following specific information about your plant:

- Common name
- Scientific name
- Plant family
- Plant description

- Place of origin
- Ecology and distribution (Where does it grow in the world? Is it wild or cultivated? Is it common or rare?)
- Cultivation practices
- Reproductive biology (flowering, pollination, and fruit or seed development)
- Traditional uses (food, medicine, fibers for clothing or paper, construction, etc.)
- Method of use (details on preparation, storage, and use)
- Safe use
- Modern pharmaceutical, nutritional, or other industrial uses
- The findings of scientific studies conducted on useful properties of the plant
- Other information (folklore, recipes, historical events associated with the plant)

Is This Plant Dangerous?

Like animals, some plant species are dangerous and are best left alone. Get to know your plant before actually working with it, or it could hurt you. Make sure that the final plant you select to study and experiment with is a safe one. Poison ivy and poison oak, for example, can cause a severe skin rash from even casual contact with any part of the plant. Stinging nettle causes a painful prickly sensation when handled. Other plants can be handled without incident but are toxic if ingested. Sassafrass (*Sassafrus albidum*), for instance, has a cancer-causing substance called safrole, and pennyroyal (*Mentha pulegium*) can cause liver damage or even death. Other plants may be unavailable or illegal to use. Good background research from reputable sources can help you make solid choices and provide safe leads for later experimentation. Unit 7 of this book has more information on finding a safe plant to study and experiment with.

In the appropriate sections of your report, you should mention whether what you learned from the literature verifies or conflicts with what you learned from your informant. This is part of the scientific process in ethnobotany. Look at the following example from Pablo's report on dragon's-blood:

> In Ecuador our people use the red sap of the sangre-de-drago (*Croton lecherii*) on wounds and cuts. In the interview the man said that it stops bleeding and is used as an antiseptic to kill bacteria and stop infection. But the article that I read said that dragon's-blood (as they call it in English) is being tested as a new medicine for viral diseases like herpes, the flu, and hepatitis. So I'm not sure if the plant kills bacteria or viruses or both.

Before we go on to describe the correct format for various types of bibliographic citations, we first want to say a few words about the relevant yet sometimes overlooked topic of copyright. When you turn to outside sources (books, journals, Web sites, and so on) for images, sounds, video clips, or long excerpts of text, you must check to see if the items you wish to use are protected by copyright. If so, it would be illegal for you to copy and reuse them (in a science fair display, a Web site, a photocopied hand-out, or a similar situation where it will be displayed or reproduced) without the written permission of the publisher, artist, or author.

In some cases, images are free to copy and reuse. However, if a source does not specifically say that this is the case, then you must ask for permission. Briefly explain what you will use the material for and how it will be presented, and ask how the author (or artist or publisher) would like to be given credit for use of the item.

On the following page are descriptions and examples of how authors of scientific papers cite their sources. If your teacher prefers a different method for any of these, you should follow that preference.

Book

Format:

Author last name, author first name. Year of publication. Full title of work. City of publication: Publisher.

Examples:

Chu, Thomas J. 1999. *The Science behind the Folklore: Investigations into Folk Healing.* New York: Folk Pharmaceutics Press.

Vargas, M. Teresita. 1999. *Herbals through Time: The Historical Development of the Pharmacopeia.* Hudson, NY: Spinderly Publications.

Journal

Format:

Author last name, author first name. Year of publication. Full title of article. Name of journal, Volume (Issue): Page numbers.

Examples:

Bender, Leela. 1999. "Chemical composition of ten popular home remedies." *Journal of the Unknown*, 28 (2): 156–162.

Al-Hamman, Ali. 1999. "Medicinal uses of black cohosh." *Herbal Medicine Chest*, 1(April/May): 48.

Web site

Format:

Author last name, author first name. Last date visited. Full title of web page. URL.

Examples:

American Association of Cardiac Surgeons, Nurses, and Health Care Workers. September 1999 (last visited). "Heart health: Twenty foods for staying fit after the bypass." URL: < www.aacsnhcw.org/h_health.html >.

Stevens, Chitra D. September 1999 (last visited). "My plant collection." URL: < www.middleschoolamerica.net/users/ ~ c_d_stev/myherbarium.cgi >.

E-mail

Format:

Author last name, author first name. Date of message. Subject. Electronic mail.

Examples:

Raychaudhury, Srinistava. 22 August1999. "Ayurvedic herbs." Electronic mail.

Gudmondsson, Sven. 10 March 2000. "Your question about roots." Electronic mail.

Interview

Format:

Informant last name, informant first name. Date of interview. Interview (or telephone interview) with author, city where interview took place [only for in-person interview].

Examples:

McMichael, Eric. 2 September 1999. Interview, Ghent, NY.

Lauren, Rachel. 16 April 2000. Telephone interview.

Unit 2 Questions for Thought

On a separate piece of paper, answer the following questions as thoroughly as possible.

1. What can ethnobotanists learn from elders and healers?

2. Where are some good places to go to find someone to interview?

3. If someone tells you something about a plant that sounds strange or unrealistic, what could you do to verify whether or not the information is true?

4. What might happen if you were not respectful to the person whom you are interviewing?

5. Think of someone or several people that you know who would be really good to interview. Who are they and why do you think they would know a lot about the useful properties of plants?

6. What characteristics should you look for when finally selecting the plant you will study?

7. Where should you go on the Internet to find reliable sources of information about the plant that you are studying?

8. Why is it important to write the paper in your own words rather than copying sections of text into your paper?

9. Read the paragraph below and write a paragraph that uses the facts or ideas presented below, but phrased in your own words:

> The World Health Organization is involved with research, which claims to evaluate the effectiveness of various types of traditional remedies. Since 80 % of the world's people already depend mainly on herbal and traditional remedies and because pharmaceutical drugs are frequently expensive and difficult to transport to many parts of the world, people should be urged by medical practitioners when possible to use traditional, local medicines which are demonstrated to work to promote good health. [Schultes and von Reis 1995, p. 27]

10. Why is it important that you synthesize information from different sources? In other words, why should you review all the information from different places (interviews, the Internet, journals and books), put all these ideas together and come up with your own ideas, questions and opinions?

11. How should a bibliographic citation of a book differ from that of the following:
    a. research journal?
    b. website?
    c. email?
    d. informant interview?

Unit 2 Activity

Field Exploration: The Ethnobotanical Interview Form

Student (interviewer) name: ________________________________ Date: ____________

Interview location (city, state, country): ________________________________

To help explain the different purposes of each section of this interview, you may want to use the paragraphs below written in italics. Put these your own words.

Student: *"In our science class we are studying ethnobotany, the relationship between people and plants. Our assignment is to interview someone who knows a lot about plants from their own culture. This information will help me choose a plant for a research project. In the first part of the interview I'll ask you a few questions about your background. In later parts I'll ask you specific questions about plants."*

Section A. Background Information about the Informant

Name: ________________________________

Current residence (city, state): ____________________

Birthplace: ________________________________

Age (estimate, if they don't want to say): ________________

Native language: ________________________________

Ethnic background or nationality: ____________________

How did informant learn about plants? (From parents? Grew up on farm? Gardening experience?) ________________________________

____

Section B. A General List of Useful Plants

Student: *"Now I would like you to think of the most useful and interesting plants that you know of, especially those that have been important in your family. I just want to make a general list."*

What plants have been used traditionally in your family (such as foods, medicine, skin or hair care, for holiday traditions, as dyes, for traditional crafts…)?

Plants mentioned Use(s)

1. ____________________ ____________________________
2. ____________________ ____________________________
3. ____________________ ____________________________

4. ____________________ ______________________________
5. ____________________ ______________________________
6. ____________________ ______________________________
7. ____________________ ______________________________
8. ____________________ ______________________________
9. ____________________ ______________________________
10. ____________________ ______________________________

Section C. A Short List of Promising Plants

Student: *"Now I want to narrow the list down to the most promising plants for research by asking you further questions."*

(Ask the following questions, or make up your own questions, to help you select two or three of the most interesting plants.)

Which of the plants you mentioned is your favorite, and why? ____________________

____

Which plant do you think is the most nutritious? ____________________

Which plant is the most effective medicine? ____________________

Which plant has been most important as part of your family tradition? ____________

Based on the above questions, select one to three of the most interesting plants for further investigation.

Plant Use(s)

1. ____________________ ______________________________
2. ____________________ ______________________________
3. ____________________ ______________________________

Section D: Detailed Information about Each Plant

Fill out the form below for each of the plants on your short list in section C. (Make several copies of this form so you can conduct a detailed interview for each plant on your short list.)

Common name(s) of plant: ______________________________

Scientific name of plant (if known): ______________________________

Is this plant safe to use? _____ If not, explain why: ______________________________

Is this plant legal to use in this country? _____ (If not, stop here and choose another plant.)

Describe plant uses in greater detail if possible:

Food: ______________________________

Plant part used: ______________________________

Method of preparation: ______________________________

____

Medicine: ______________________________

Plant part used: ______________________________

Method of preparation: ______________________________

____

Cosmetic (including soaps, shampoos, skin tonics): ______________________________

Plant part used: ______________________________

Method of preparation: ______________________________

____

Dyes: ______________________________

Plant part used: ______________________________

Method of preparation: ______________________________

____

Weaving, rope-making, or other use of fibers: ______________________________

Plant part used: ______________________________

Method of preparation: ______________________________

____

Building materials: ______________________________

Plant part used: ______________________________

Method of preparation: ______________________________

____

Ornamental or traditional holiday use: ______________________________

Plant part used: ______________________________

Method of preparation: ______________________________

____

Pesticides or other uses: ______________________________

Plant part used: ______________________________

Method of preparation: ______________________________

____

Important recipes: ______________________________

Plant part used: ______________________________

Method of preparation: ______________________________

____

Does this plant have any special cultural importance in your family? _____ If so, please explain: ______________________________

____

Do you have any interesting stories about this plant? ______________________________

____

____

____

____

____

How to find the plant:
Does it grow here (in the area, region, town, etc.)? ______

If yes, is it wild or cultivated? __________________

If wild, where can it be collected? __________________

If it does not grow here, where does it grow? __________________

Is it wild or cultivated in its native habitat? __________________

Can you buy it in the store? _____

If yes, which store(s)? __________________

Is it sold fresh, dried or both? ____________

If you cannot buy it in the store, how do you get it? ________________

Do you know where I could find any photos or drawings of this plant? ____________

____

____

Do you know of any books that describe this plant? ______________________________

____

____

Any other relevant information? ______________________________

____

____

Remember to thank the informant for his or her time and help with this interview.

Checklist of Activities for Report

The following checklist will help you complete all parts of your report. As you complete each item, place a checkmark in the space to the left of it. If you see blank spaces as you look down your list, you will need to return to and complete the missing item(s).

Your teacher can also use this checklist as a grading tool, by assigning one of the following numbers in the space to the right of each item: a **4** for "excellent," a **3** for "good," a **2** for "needs improvement," a **1** for "very poor," and a **0** for "not done at all." Your teacher will then add the total score and divide it by the total possible score (80) to determine a final grade for the project.

Name: ______________________ Date: ______________________

Topic of paper: ______________________ Class: ______________________

Did I . . . Score (to be filled in by your teacher)

_____ 1. Find someone to interview? _____

_____ 2. Conduct the interview and produce a short list of interesting plants? _____

_____ 3. Check general library and Internet references on short list of plants? _____

_____ 4. Make final selection of a plant for in-depth study? _____

_____ 5. Prepare detailed outline of the report? _____

_____ 6. Conduct further library research? _____

_____ 7. Download relevant Internet information on the plant from well-respected botanical sites? _____

_____ 8. Take notes and cite sources on index cards? _____

_____ 9. Write the first draft of the paper? _____

_____ 10. Include most of the suggested aspects of an ethnobotanical report (such as common name, scientific name, family, origin, history, cultivation, ecology, traditional uses, modern uses, scientific studies, and folklore about the plant)? _____

_____ 11. Rewrite and revise the paper, making sure that the writing is in my own words? _____

_____ 12. Check the grammar and spelling and correct the errors? _____

_____ 13. Type the paper (using a 12 point font, double-spaced) or, if I don't have access to a typewriter or computer, print the text neatly? _____

_____ 14. Make the body of the paper at least 5 to 10 pages long when typed? _____

_____ 15. Prepare an attractive cover page with the title, my name, the date, my teacher's name, and the class title? _____

_____ 16. Prepare a table of contents? _____

_____ 17. Add attractive drawings, photographs, graphs, and/or charts to the paper? _____

_____ 18. Write the introduction and summary, adding my own evaluation and opinion based on my background research? _____

_____ 19. Write the bibliography, with sources in the correct format? _____

_____ 20. Make sure that I cite at least 4 to 6 sources in the bibliography and that the sources include the informant interview, the literature, and Internet sources used? _____

Total score _____

Total score divided by 80 = _____%

Unit 3
How to Preserve, Represent, and Reproduce Your Plant

Introduction

Imagine that you have discovered a new plant that may have many amazing uses. You would surely want to preserve this plant for further study, and draw or photograph it so you could let others know about its benefits. Most importantly, you would want to propagate it for further distribution. That's what this chapter is about — exploring different methods of preserving, representing, and reproducing your study plant.

In this unit you will learn to make herbarium specimens, to illustrate and photograph your plant (as the West Roxbury High School students did in the poster shown above), and to transfer botanical images.into the computer. You will also learn how to dry plants and make extracts and tinctures from them. These are all methods that professional scientists use to preserve and represent plants, and these techniques will help you communicate your own findings in a way that other people will easily understand. Finally, you will try to reproduce your plant in the lab from seeds and/or cuttings.

Plant Conservation

Broadly speaking, plants can be preserved in two ways: They can be protected in their natural habitat or agricultural setting, or removed and safeguarded in specialized institutions such as botanical gardens and seed banks. The conservation role played by these botanical institutions is similar to that offered by zoos for endangered animals. Since the late 1980s, botanical gardens have collaborated with an international conservation network to make sure that the rarest or most endangered plants are prioritized for propagation and, ultimately, for reintroduction to their native habitats. Botanical gardens also preserve and reproduce plants in other ways that apply directly to the activities in this unit.

Botanical Gardens

Many of the world's active botanical gardens were first established by European colonial powers hundreds of years ago as a repository for new and promising plants gleaned from their newly conquered territories. Today, however, most botanical gardens have redirected their mission toward education and species preservation. Combined, the world's 1,600+ botanical gardens house representative specimens of perhaps 25% of the world's flora.

Botanical gardens usually maintain two main types of plant collections: first, a diverse collection of living plants; and second, an *herbarium*, a collection of dried, pressed plant specimens. In addition, some botanical gardens, as well as many agricultural institutions, maintain *gene banks* which focus largely on saving seeds of crop varieties and their wild relatives. Gene banks play an important role in the conservation of plant genetic diversity.

A Safety Net for Endangered Plants

Twenty-eight of the leading botanical gardens and arboreta in the United States form a network called The Center for Plant Conservation, or the CPC. Together this network houses the National Collection of Endangered Plants — 549 of the most endangered species in the United States. Through their programs in research, education, and conservation, the CPC is working to prevent the extinction of our national botanical treasures. A survey in 1988 showed that three-quarters of all the endangered plants in the United States are found in five areas: Hawaii, California, Texas, Florida, and Puerto Rico. These states have since then been given high priority for conservation efforts by the CPC.

Living Plant Collections

Living collections of woody and herbaceous (non-woody) plants held in botanical gardens are usually maintained in specialized outdoor and indoor gardens. An *arboretum* is a collection of different kinds of trees and woody shrubs, often growing in a forest or park-like setting. In cool temperate climates, greenhouses with controlled interior climates protect collections of tropical and desert plants such as orchids, palms, and cacti. Collections of aquatic plants like water lilies may be safeguarded in natural or man-made ponds.

Many of the prominent botanical gardens of the world are distinguished by their unique collections of living plants. The Fairchild Botanical Garden in Florida maintains one of the world's most complete collections of palms and cycads. The Desert Botanical Garden in Phoenix, Arizona, contains a renowned collection of desert succulents including the extraordinary Saguaro cactus. The Arnold Arboretum, near Boston, is home to one of the most diverse outdoor collections of hardy temperate-zone trees, a specialty being trees from China. And visitors to The New York Botanical Garden can walk through the plant life of different biomes (such as the Palms of the Americas gallery, at left, along with lowland tropical forest, desert, and other biomes) in the renowned Enid A. Haupt Conservatory. But some of these botanical gardens are even better known, among scientists, for their herbarium collections.

photo courtesy of The New York Botanical Garden

Herbarium Collections

When plant scientists go into the field, whether to search for new species or to conduct floristic or ethnobotanical surveys, they make plant collections. These plant samples are pressed (see the diagram of a typical plant press, top of the facing page), dried, and deposited in an *herbarium* — a collection of preserved plant specimens — so that they can be studied and reexamined carefully by the collector or other scientists, often far from their original habitat.

These preserved plant samples, known as *herbarium specimens* or *voucher specimens,* serve as permanent records of botanical information. For ethnobotanists, these specimens form an important link between folk knowledge and scientific knowledge. Because the common names of plants vary from culture to culture and region to region, we must rely on the proper identification of a plant specimen to establish its scientific name.

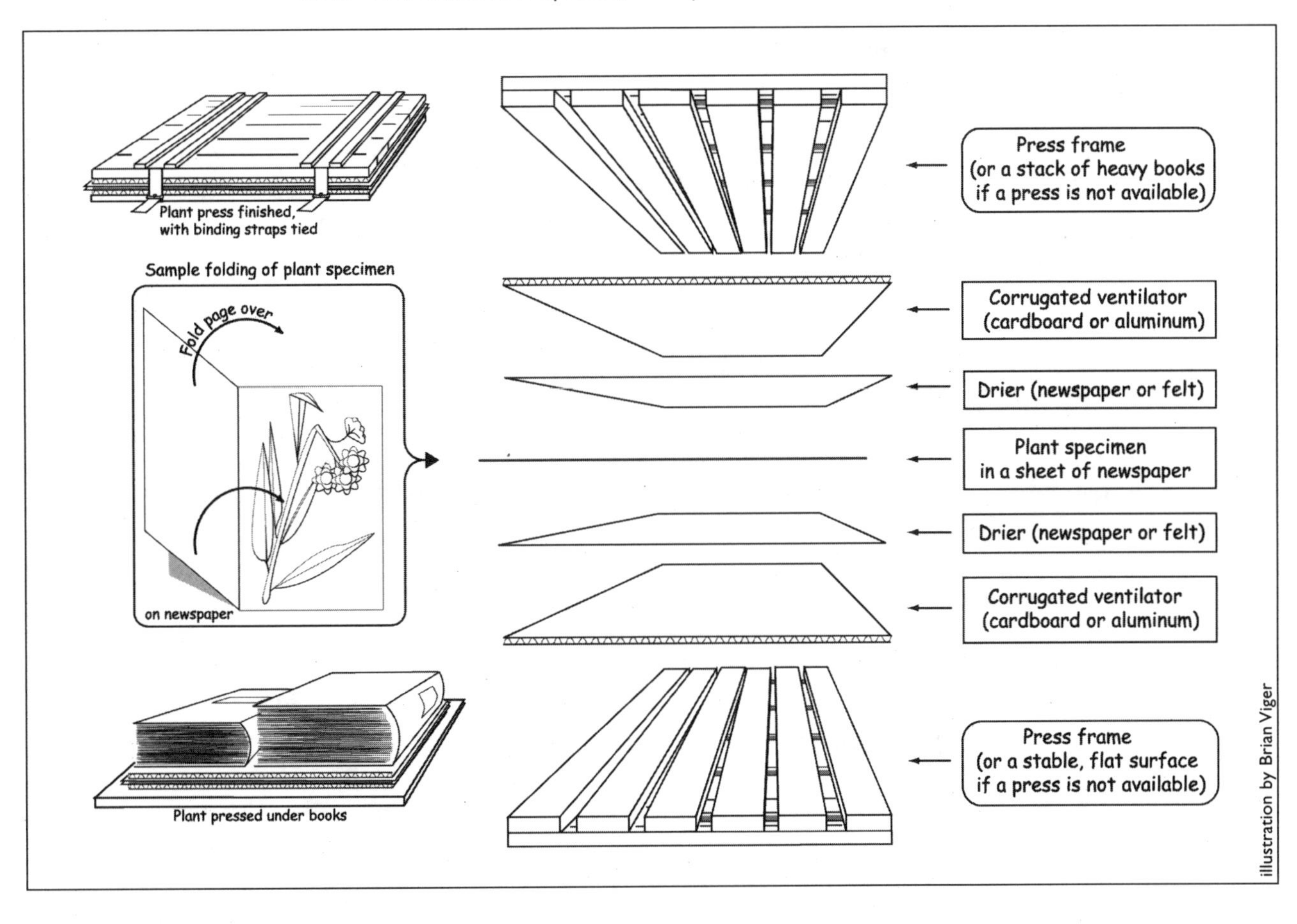

illustration by Brian Viger

Without proper voucher specimens, the ethnobotanical work conducted in one region of the world cannot be compared to another, nor can it be verified by other scientists.

Botanists rely on herbarium specimens to identify and classify new species. When properly prepared and stored, these specimens can last for hundreds of years. When a botanist describes a new species for the first time, a *type specimen* is selected and deposited in an herbarium, to represent that species. The New York Botanical Garden contains approximately 6 million herbarium specimens, of which about 125,000 are type specimens.

An herbarium specimen (such as the one at right, showing *Macleania pentaptera*) consists of a well-preserved plant section containing vegetative parts (leaves and a portion of the stem) and reproductive parts (flowers and/or fruits) mounted on a piece of stiff paper that typically measures 11.5" × 16.5". A label in the corner of the page gives a description of the living plant's appearance, the place and date the specimen was collected, the collector's name, and other relevant information. (One of the activities at the end of this unit will describe how to collect a plant and make it into an herbarium specimen.) After preparing an herbarium specimen, a scientist needs to identify the plant by its scientific name.

photograph courtesy of The New York Botanical Garden Herbarium

Plant Identification

Field Guides and Dichotomous Keys

Most scientists, naturalists, gardeners, and students use field guides or other plant identification books to determine a plant's name. Field guides describe plants using both text and clear line drawings or photographs. These books are generally written to help the reader identify a plant using common field characteristics, and don't usually require knowledge of complex botanical terminology. They may focus on plants of a particular region, or on specific groups of plants such as ferns, grasses, weeds, wild edible plants, or trees. In spite of their diverse scope and complexity, most field guides are united by their use of dichotomous keys.

A *dichotomous key* is a step-by-step guide to help you identify your plant based on a series of questions with only two possible answers to choose from for each question. These questions lead you along a trail of clues in search of your plant's identity.

Below is an example of a very simple dichotomous key with nine couplets. Using the key, you can identify the tree species for each of the ten leaves shown below. Start by reading couplet 1 and determine if the leaf you are trying to identify is described better by 1a or 1b. The *lead* (the number or name at the end of the dotted line after each part of the couplet) will tell you what couplet number to go to next so you can narrow down your search. If it says go to 6, then read couplet 6 and decide if the leaf is best described by 6a or 6b. Keep following the key in this way until it takes you to the name of the species you are looking for. When you find the name of the plant, write it underneath the leaf in the space provided. You can double-check your identification by looking up the species in a field guide.

Here is an example of how to use the key for the first leaf shown (upper left corner). Reading couplet 1, since the leaves are not evergreen (1a) but are broad and deciduous (1b), the lead directs you to couplet 6. The leaves are not compound or divided into leaflets (6a) so you should go on to 6b, simple leaves, which directs you to couplet 7. There you should select 7b (toothed leaves), which takes you to couplet 8. At 8 the correct selection is 8a (long slender leaves that droop), and the lead tells you that the plant is weeping willow, the scientific name of which is *Salix babylonica*. See if you can identify the other nine species by using the key. Write the name on the line below each leaf.

A key to the leaves of some common trees of the northeastern United States

1a. Leaves are evergreen, thin, needle-like 2
1b. Leaves are broad, deciduous 6

2a. Needles are more than one inch long, in clusters 3
2b. Needles are one-half inch or less 4

3a. Needles are in clusters of 3 .... Pitch pine *(Pinus rigida)*
3b. Needles are in clusters of 5 .... Eastern white pine *(Pinus strobus)*

4a. Needles are scale-like, sharp, covering the twigs .... Eastern red cedar *(Juniperus virginiana)*
4b. Needles protrude from the twigs 5

5a. Needles are flat, rounded tips in 2 rows along twig .... Eastern hemlock *(Tsuga canadensis)*
5b. Needles are in a whorl around the stem .... White spruce *(Picea glauca)*

6a. Leaves are compound and divided into 7 leaflets .... White ash *(Fraxinus americana)*
6b. Leaves are simple 7

7a. Leaves are rounded and have 7 to 9 lobes .... White oak *(Quercus alba)*
7b. Leaves are toothed 8

8a. Leaves droop and are long and slender .... Weeping willow *(Salix babylonica)*
8b. Leaves are less than twice as long as broad 9

9a. Leaves have an elliptical shape .... American beech *(Fagus grandifolia)*
9b. Leaves are toothed and lobed .... Sugar maple *(Acer Saccharum)*

If you have not already identified your study plant by its scientific name, we encourage you to do so using Mabberley's *Plant-Book* (D. J. Mabberley, *The Plant-Book: A Portable Dictionary of the Higher Plants.* New York: Cambridge University Press, 1989) or other reference works (see Appendix A).

Representing Your Plant in Pictures

By drawing and/or photographing your plant, you can learn about its distinguishing characteristics and anatomy, in addition to showing other people what your plant looks like. While it is best to find a living specimen that includes most of the important parts (stem, leaf, flower, root, fruit, seed), this is not always possible. If your plant does not grow in the area where you live, or if it is not growing during the season you need to draw it, you may not be able to draw a living specimen. In that case, it is permissible to use a photograph or drawing in a book as a reference for your own illustration.

Plant Illustration

The clear, black-and-white line drawings typically found in plant identification books arise from a rich tradition of botanical illustration. Although botanical art dates from earliest civilization and illustrated herbals are known from the first century A.D., true botanical illustration dates from the Renaissance. Prior to that, most illustrations found in the ancient herbals were drawn from earlier illustrations rather than from the actual plants themselves. Botanical art truly began to flourish during the Renaissance, when illustrators began to focus on the characteristics that distinguished one plant from another, as the art of illustration became more scientific.

There are two main advantages of illustration over photography for plant identification. First, the drawing can focus solely on the plant without involving distracting background colors and objects. Second, details are usually clearer in a drawing, where the line and shape of the object are emphasized, making it sharper for observation and identification. For instance, with an illustration you can distinguish the ways in which petals overlap or the types of hairs that cover a leaf. An artist can emphasize the parts of the plant most important to identification. Likewise, botanical illustration also serves as an important teaching and research tool; it is a way for a scientist to record observations. The act of drawing forces people to slow down and carefully notice what they see before them.

One of the activities at the end of this unit will describe the standard techniques of botanical illustration so you can draw your study plant. By having students draw what they see, a teacher can check a student's understanding of plant anatomy. In this unit, plant illustration can be both an exercise in careful scientific observation and an expression of individual creativity, as is evident from the group of students' illustrations at the start of this unit.

Plant Photography

Today most field guides include photos as a way of identifying plants. Unlike illustration, however, a photograph contains no lines; instead, a pattern of grays and colors create the image. The word "photography" comes from Greek words meaning "light drawing." While lines are essential for discerning particular details needed for plant identification, photos provide a complementary means of botanical image presentation in books or reports. In the lab activity at the end of the book you will find guidelines for photographing your plant.

Computer Graphics

You can put your plant images into the computer and use them to publish reports, prepare poster presentations, put together digital slide shows, or create web pages. While you can make a drawing directly on the computer using the proper software, it is more common to transfer images from hard copy to a digital format. Three methods exist to accomplish this: First, you can copy the drawings or photographs by using a scanner to transfer the images into the computer. Second, you can take photographs of your plant using a digital camera that transfers the images from the camera's internal memory to the computer. Third, you can take photographs using your regular camera and have the film processor put them on a photo CD.

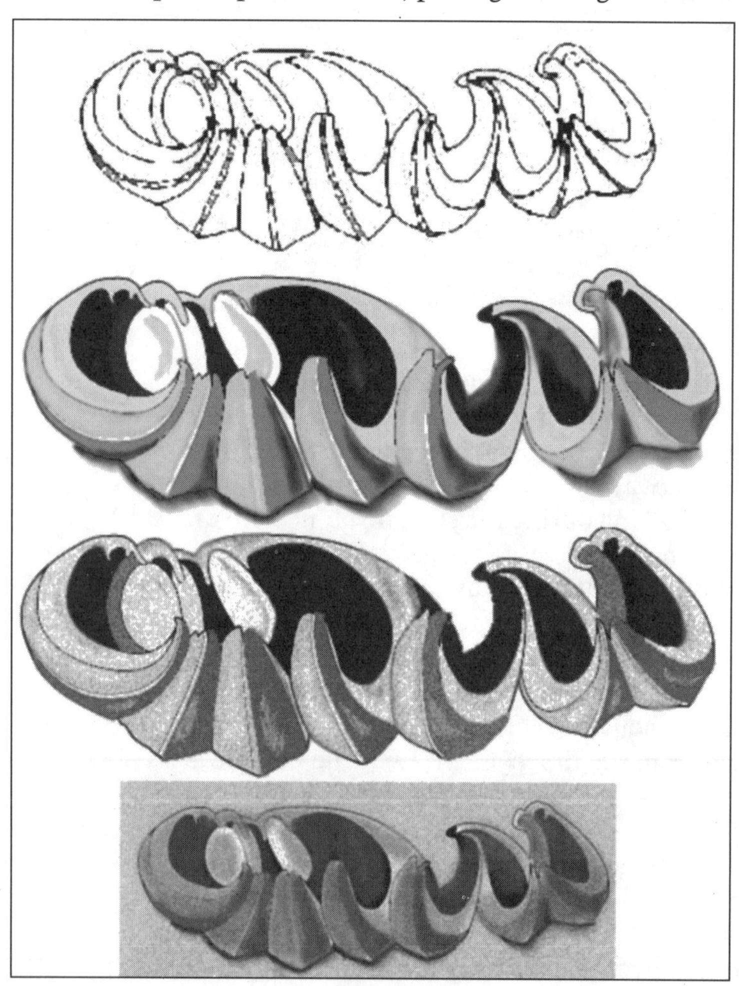

Once inside the computer, the plant images can be modified by using a graphics program. They can be scaled to any size you wish, lightened, darkened, or cropped. To the left you can see how the seed pod of *Hura polyandra* has been manipulated from the original line drawing. In the activities at the end of this chapter you are encouraged to transfer your illustrations and photographs onto the computer and alter these images to best fit the needs of your report or presentation materials.

Plant Propagation

Perhaps the most important technique for preserving your plant involves growing it. What value is there in knowing a plant's useful properties if it cannot be reproduced? In this section we will discuss two general, easy methods of propagating plants.

The art of growing a new plant, *plant propagation*, falls into two categories: sexual and asexual reproduction. *Sexual reproduction* entails growing a plant from seed. *Asexual reproduction* involves growing a new plant from a vegetative part of the parent.

Seeds usually result from pollination between flowers — that is, the transfer of pollen from the male part (stamen) of one flower to the female part (pistil) of another flower facilitated by wind, insects, or even birds or small mammals. Hence, the offspring or seeds will contain a different combination of genetic information from that of the original parents. This act of recombination maintains the foundations of genetic diversity, the basis for future adaptation to environmental change.

Growing a plant from seed is the most common propagation method. Some seeds may require special treatment before they will germinate such as *scarification* (a scratching of the seed coat; this may take place as seeds pass through the digestive tract of a bird), a requisite cold period (plants evolved a cold requirement to protect seeds from germinating before winter has passed), and, for some fire-dependent species, germination occurs only after a burn has taken place. However, most seeds germinate easily when supplied with their basic needs: water, oxygen, soil, and an appropriate temperature.

Asexual or vegetative propagation produces a young plant that is genetically identical to its parent. The offspring or new plants result from a type of "cloning" — the production of a new plant from a piece of an older one. This ability is often related to extreme habitats where seedlings would have a difficult time surviving. Asexual reproduction enables a plant to develop without having to go through the vulnerable phases of germination and establishment.

The three most common techniques of asexual plant reproduction include propagation by (a) *division*, (b) *cuttings*, and (c) *layering*. Many other propagation techniques exist that will not be discussed here, including budding and grafting, which are more difficult to master, and tissue culture, in which plants are reproduced from just a few plant cells under controlled laboratory conditions.

Division involves separating a single plant into two or more groups. Clump-forming plants with underground bulbs or rhizomes are easily divided and replanted. Examples of plants best propagated by division include aloe, aster, banana, daffodil, day-lily, garlic, and iris.

Using *cuttings* to propagate plants is also a simple process and successful for many species. A stem cutting that includes leaves and buds will often root when placed in moist soil or a soil-sand mixture. Cuttings may also root when placed in a vase of water. Some plants also root from leaf or root cuttings. Some plants that can be propagated from cuttings include begonia, coleus, comelina, English ivy, geranium, jade plant, watercress, and potato.

Layering involves stimulating roots to grow from a stem or branch while it is still attached to the parent plant. This can be as simple as bending a stem to the ground and covering it with some compost or soil. Once the roots have formed, the stem is cut off and planted on its own. Among plants that can be propagated by layering are dogwood, forsythia, raspberry, rubber plant, spider plant, and strawberry.

Once you propagate your plant successfully, you need to learn how to keep it growing and healthy. The information you gathered during your background research may be helpful to you now as you try to grow your plant. Try and provide your plant with its particular growth requirements as much as possible. In temperate zones, for example, tropical plants may be grown successfully in the home near a sunny window, or tropical perennials might be grown outside as an annual during the summer months. Consult books, gardeners, extension agents or perhaps the person you interviewed for advice on raising your plant. Finally, the best way to learn is to try and grow it yourself, learning from your successes and failures.

Sometimes the growing conditions required for your particular plant make it impossible to grow it in the area where you live. Some plants require very exacting conditions. So don't feel discouraged if you are unable to grow your plant. If you try your best you will learn something, regardless of whether or not the plant grew successfully.

Unit 3 Questions for Thought

On a separate piece of paper, answer the following questions as thoroughly as possible.

1. What are the advantages of conserving plants in each of the following institutions:
   a. seed bank?
   b. botanical garden?
   c. arboretum?

2. What are the advantages of conserving plants and other species in their natural habitat?

3. What does "CPC" stand for, and what does this organization do?

4. Would preserving living plant specimens protect species better than saving the seeds in a seed bank? Give reasons for your answer.

5. What is the purpose of an herbarium collection?

6. What are some of the ways that plant essences can be preserved? What are these products used for?

7. What is the purpose of field guides and dichotomous keys?

8. Why are hand-drawn plant illustrations still sometimes preferred over photographs?

9. What are some of the differences between sexual and asexual reproduction in plants?

10. Which plants or types of plants are best started from each of the following?
    a. seed
    b. root cuttings
    c. stem or leaf cuttings
    d. division
    e. layering

Unit 3 Activities

Field Exploration: Make an Herbarium Specimen of Your Plant

In this activity you will make an herbarium specimen from your study plant. If possible, collect a live specimen of the species you have chosen to study as well as several other species of plants growing around your neighborhood or schoolyard which are interesting because of their abundance, beauty, or usefulness. You can use the skills you gain in preserving plant specimens to create your own or a class "field herbarium" with mounted specimens held in a large three-ring binder. Each time you encounter a new species, you can enter a new page in your herbarium.

Materials needed
Fresh plant specimens
Plant press, or several heavy "dictionary-sized" books
Sheets of newspaper
Sheets of felt, or additional newspapers
Sheets of corrugated cardboard, cut to about 12" x 18"
Stiff sheets of paper, about 11.5" x 16.5"
Glue, tape, or needle and thread
Label for your herbarium sheet
Local or regional field guides for identifying your plant collections

Directions

1. Collect fresh plant specimens that contain a portion of stem with leaves intact, and whenever possible, reproductive structures (flowers and/or fruits). With small plants, the roots can also be collected.

2. Specimens should be placed neatly between folded sheets of newspaper for drying. Plants may need to be cut and the leaves or stems folded, as shown in the illustrations on page 31. Flowers should be pressed open. Fleshy fruits or tubers can be sliced thinly to press and dry.

3. While most botanists use a wooden plant press to dry specimens between newspapers and cardboard (see illustration on page 31), students can improvise by pressing plants in newspapers under any flat and heavy object. For example, specimens pressed in newspapers may be placed under books or boards weighted with something heavy. The pressed plants should be placed in a warm, dry, preferably sunny place in the school or home. Ideally they should be checked every day or two, and damp newspapers must be replaced with dry ones to avoid damage by fungi or insects. It should take approximately one week for plants to become dry enough to be mounted.

4. Mount the dried plants on stiff sheets of paper using glue, tape, or needle and thread. Make sure each plant is labeled with the following information in the lower right-hand corner (use a field guide to help you identify the plant and learn about its important characteristics):

    Name of collector (your name)
    Date of collection
    Location where collected
    Type of habitat
    Common name of plant
    Scientific name of plant
    Plant family
    Brief description of identifying characteristics
    Plant use or other points of interest

If you do not wish to make your own specimen label, you can photocopy the one that appears at the bottom of this page. (It is **not** recommended that you cut it directly out of this book.) Once you've filled in the information, you can cut out the label and attach it to your specimen.

5. Optional: Although this is never done in a professional herbarium collection, you might consider having your specimens laminated so they will be better able to withstand a lot of handling. Another technique is simply to minimize handling by scanning the finished specimens into the computer and creating a "digital herbarium."

Additional Activity

Create a class field guide or field herbarium using dried plant specimens on a particular topic of interest. Below are a few suggestions:

- Medicinal plants around our school
- The wetland plants of our community
- Weeds in the neighborhood
- Plants to attract birds and butterflies
- Wild edible plants in our community
- Plants used by the Native Americans in our region

Collector's name:

Date collected:

Location where collected:

Habitat:

Common name(s) of plant:

Scientific name: Family:

Identifying characteristics:

Uses or other points of interest:

Field Exploration: Create a Botanical Illustration

Materials needed

Drawing paper; drawing notebook
Pencil, sharpener, eraser, ruler, compass
Clipboard or drawing board to use as a hard surface
Optional: watercolors; pen and ink; hand lens or dissecting scope; dissecting tools.

Directions

1. Find a well-lit place to draw. In botanical illustration, the light always shines on the plant from the upper left hand side. This is important to remember when adding shading.

2. Set up your fresh plant material as you would like to draw it. Draw the entire plant first; this is called the *habit sketch*. Begin drawing some guidelines on the paper so that the plant image will fit. Separate details and enlargements can be added in the margins around the habit sketch (a standard practice in botanical illustration). You want to show the position of the leaves and pattern of the veins. A close-up of the flowers or fruits should be included that show key details such as petal number and position; stamens, pistil, etc. Keep in mind that you want to emphasize the characteristics that help identify your plant species. Look at plant illustrations in books for layout ideas.

3. Measure the length and width of the leaves. Try to draw your plant in accurate scale. Keep your ruler or compass handy, and make frequent measurements as necessary.

4. Use light lines until you are satisfied with your illustration. When you are ready, trace them over in dark pencil or ink. You may want to trace your final illustration from your initial draft.

Additional Activities

A. Try drawing your plant at a different angle, or from farther away.

B. Label the parts of the plant in one of your drawings. Use a plant anatomy book so that you become more familiar with the parts and structure of your plant.

C. Illustrate another plant that you find interesting. Try drawing trees or shrubs in an outdoor setting.

D. Consult the recommended plant illustration books and try your hand at shading, or try using colored pencils or watercolors.

Field Exploration: Photograph Your Plant

Materials needed

Camera with film or digital camera
400 or 200 ASA/ISO film (if using outdoor color film)
Tripod (optional)
Flash attachment (optional)
Black velvet or terrycloth material to use as backdrop for plant (optional)

Directions

1. Look for visually appealing angles to take your photograph. Look for attractive patterns of light falling on the plant. Take several shots of the same plant, moving around it, closer and farther away and, if possible, from above and below it.

2. If you want a photograph without the distraction of other plants or background, place the black material so that it is behind the plant, relative to your intended shooting angle. (You might need to ask someone to hold up the backdrop for you while you shoot the picture.)

3. If you are outside shooting on a windy day and you have a camera that has an adjustable shutter speed, adjust that speed to 1/60 of a second, or faster, to get a crisp image. A slower shutter speed will create a blurred or fluttering effect.

4. If you are shooting indoors, keep the light source to the left (this is standard in scientific illustration and photography).

Additional Activities

If you put your images into the computer by way of a digital camera, scanner, or photo CD, there are many ways those images can be used. You could make a digital poster or digital slide show about your project. You can print out the image to use with your final report or as part of the final science project and poster. You may also be able to put your digital images on your class's or school's web site for everyone to see.

Laboratory Activity: Propagate Your Plant

The materials and guidelines on this page apply to parts A and B of this activity, which you will find on the following two pages. As you will see, each part has its own specific directions and analysis that you will need to complete.

Materials needed

Seeds and living plants (or fresh stems, roots, or branches) of the species you have chosen to study
Sterile potting soil
Planting containers
Wooden or plastic labels
Waterproof permanent marker
Knives or scissors for cutting plant stems
Small watering can or spray bottles
Optional: Rooting hormone (substance that stimulates root growth in cuttings – sold in nursery supply stores)
Optional: Plastic soda pop bottles or plastic bags to create miniature "greenhouses"

Guidelines

1. Sometimes it is not possible to grow your selected species if it is an exotic plant for which seeds or live plant cuttings are not available. In such a case, you can try this activity with a related or totally different species that is more easily obtainable.

2. Label containers. In this exercise you will be planting seeds or vegetative plant parts in several different containers. Each container should have a planting label that includes your name, date, plant name, and propagation method. If more than one individual is planted in a pot, make note of the number planted.

3. Keep records. Write down your methods and observations for each planting exercise in your lab book. Keep notes on plant growth over the next several weeks and possibly months. These experimental results will form part of your final report on your study plant. Your objective is to discover the most efficient way propagate your plant.

A. Propagate Your Plant from Seeds

Name: ______________________________ Date: ________________

Directions

1. Plant seeds from your chosen study plant. If you cannot get seeds for that plant, see if there is a related species whose seeds you can obtain. In general, seeds should be planted in moist soil at a depth that is about two to four times their diameter. For very small seeds, lay them on top of the soil and sprinkle a light layer of fine soil on top.

2. Gently water the soil so as not to disturb the seeds.

3. Cover the container loosely with a plastic bag or other cover to retain moisture until the seeds germinate. (The inverted bottom half of a soda bottle makes a nice greenhouse cover.)

4. Record your observations in the following Analysis section.

Analysis

What did your background research say about how to propagate your plant from seed?

____

____

____

Planting techniques used: ____

____

Number of seeds planted: ___________

Care given: ____

Germination date: ____________ Number of seeds that germinated: _______

Results: ____

____

____

B. Propagate Your Plant through Vegetative Reproduction

Name: ______________________________ Date: ________________

Directions

1. Find out which of the three following techniques can be used for propagating your study plant, and follow the directions below for the preferred method of propagation.

Division: Separate the plant at the base into several plants and put each in a separate container of soil. (For dividing garlic bulbs, separate the "bulblets" and plant each one like a seed with pointed end up.) Water.

Cuttings. Cut a healthy section of stem that contains a few leaves and buds or growing tip off the mother plant. Dip the stem base in rooting hormone and plant the stem base in moist soil. (Several inches of stem should be buried.) Press the soil firmly around the stem. Water gently. This method can also be tried using a piece of root.

Layering. Gently bend a low-growing branch so that it makes contact with loosened soil. At the point of contact, cut away a one- or two-inch-long strip of bark and apply some rooting hormone to the wound. Secure the branch to the soil with a hook or stake, so that the wound is about an inch below the surface of the soil.

2. Record your observations in the following Analysis section.

Analysis

What did your background research say about how to propagate your plant vegetatively?

____

____

Propagation technique: ____

Number of seeds/cuttings planted: ____________

Care given: ____

Date when plants showed initial signs of growth: ______________

Number of cuttings/plants successfully propagated: ______________

Results: ____

____

____

Compare your results for propagating your plant from seed and vegetatively. Which is the better way to propagate your plant, and why? ________________________

____

Additional Activities

1. Learn about different types of germination: epigeal and hypogeal. Identify what types of germination your plants display.

2. Learn about the techniques of budding and grafting. Try these techniques in the lab or at home. Learn about tissue culture and try it with the help of your science teacher.

3. Make field observations on reproduction techniques that weeds employ. Discuss why weeds are so successful.

Laboratory Activities: Preserve Your Plant's Essence for Later Use in Tests

The following extract, dried material, and tincture are preparations of your plant that will be used in subsequent laboratory activities in this book. The dried material and tincture need to be made only once; however, it is suggested that you make a new extract every three or four days, as it doesn't last as well. The extract and tincture should be refrigerated. **None of these preparations should ever be ingested.**

A. Dry Your Plant

Materials needed

Material from your plant
String
Paper towels
Mortar and pestle
Sterile glass jar (preferably tinted glass) with cover and label
Optional: Oven, drying machine, or wooden frame with wire mesh

Directions

1. No matter which of the following methods you use, you will know that your plant is "dry" when it is brittle and crunchy to the touch. (If you do not have an oven, a drying machine, or a frame with mesh, skip this step and go directly to step 2.) Turn on the oven to 120°F and place the plants on an oven rack for one to six hours, until the plants are dry. If you are using a drying machine, follow the manufacturer's instructions for that machine. If you are using a frame with mesh, lay the plants on the mesh in a single layer and for about a week, until they are dry. Proceed to step 3, skipping step 2.

2. Take a small bunch of stems (no thicker than your thumb) and tie it together with a string (if your plant has leaves, seeds, and/or flowers attached, be sure to include them). You can make several bunches if you have enough of your plant to do so. Attach the string to something so that your plant bundle(s) hang upside down in a place where air can circulate around them. Leave them there for one or two weeks, until they are dry.

3. To dry the roots of your plant, wash them vigorously and chop them into small pieces. Lay the pieces on a plate lined with a paper towel and pat them dry with another paper towel. Small berries can also be dried in this way. If the towels become wet or moist, replace them with dry towels. (Do not try to to dry fleshy or juicy fruits or stems, however, as they tend to mold easily.)

4. When the plant material is dry, crush it in a mortar and pestle. Pour the crushed material into the jar and seal it. Clearly label the jar with the contents, your name, the date of preparation, and the phrase "Warning: Do Not Ingest."

B. Make an Extract of Your Plant

Materials needed

Your plant material, preferably fresh
Distilled water (or spring water)
Enamel or stainless steel saucepan
Coffee filter, tied with string
Hot plate or stovetop
Sterile glass bottle (preferably tinted glass) with cover and label
Metric scales

Directions

1. Put 2 grams of dried plant material or 4 grams of fresh plant material into the coffee filter. Gather and tie the top of the filter, to form a "tea bag."

2. Put the tea bag in the sterile jar and clearly label the jar with your name, the plant name, the date of preparation, and the phrase "Warning: Do Not Ingest."

3. Bring the water to a rolling boil in a pan. Using a potholder to handle the pan, carefully pour the water into the jar until it is full. (Note: Your teacher may prefer to do this for you.)

4. Cover the jar and let it sit for an hour so the water cools and the tea brews. Swirl the jar once or twice during that time.

5. Remove the tea bag, seal the jar, and refrigerate.

Note: The tea should be replaced every two to four days and refrigerated when not in use.

C. Make a Tincture of Your Plant

Materials needed

Plant part you want to test (brightly colored flowers, fruits, roots, or leaves work well)
Mortar and pestle made of glass or porcelain
Rubbing alcohol
Pan
Hot plate or stovetop
100 ml beaker
Scale
Coffee filter or sieve
Small sterile glass bottle or jar (preferably tinted glass) with cover and label

Directions

1. With the mortar and pestle, grind the plant part you want to test. If you are using fresh plant material, grind 10 grams; if you are using dried material, grind 5 g.

2. Heat the rubbing alcohol in the pan over the stove until it reaches a boil.

3. Pour 50 milliliters of the hot rubbing alcohol into the beaker. While it is still hot, pour the 10 g of plant material into the beaker and grind the plant some more with the pestle. This will release the plant pigments into the alcohol solution.

4. Separate the tincture from the leftover plant solids with a filter or sieve. Discard the solids. Store the tincture in a glass bottle or jar, sealed securely so that the tincture doesn't evaporate. Clearly label the bottle with the contents, your name, the date of preparation, and the words "Warning: Do Not Ingest."

Unit 4
Testing the Nutritional Properties of Your Plant

Introduction

The use of plants as food is the most fundamental link between people and the botanical world. The study of plant foods in traditional diets forms a central part of ethnobotanical science. In this unit you will test your plant for its water content and for the presence of five types of nutrients essential to the human life: carbohydrates, protein, fats, vitamins, and minerals. (You can expand on these experiments by using the same testing procedures to test plants of the same species grown under different conditions — for instance, comparing the vitamin C content of tomatoes grown in poor sandy soil with those grown in rich loamy soil.) By knowing the nutrient composition of your plant, you may better understand its current cultural use or come up with some new potential uses of your own. **Important: Never ingest any plant that is not already in common use as a food.**

The Essential Nutrients in Plant Foods

Our bodies need food to fuel energy, growth, repair, and reproduction. These materials in food that we need are called *nutrients* and are of two types: *macronutrients* and *micronutrients*. Macronutrients, needed in larger amounts, include carbohydrates, fats, and protein. Micronutrients, required in only small amounts, include vitamins and minerals. Each of these nutrients and its functions are described briefly below.

Carbohydrates

Carbohydrates are the most abundant organic compounds in nature; they all contain carbon combined with hydrogen and oxygen in the proportions of one carbon atom to two hydrogen atoms to one oxygen atom (for example, the most simple sugar, glucose, has the chemical formula $C_6H_{12}O_6$ — for every carbon atom there are two hydrogen atoms and one oxygen atom). They are the body's main source of energy or calories. Starches and sugars are carbohydrates the body can digest, absorb into the bloodstream, and use for energy. Fiber is any carbohydrate that passes through the body unchanged and is not metabolized. Hence, while fiber provides important benefits to the body, it is not a nutrient.

Carbohydrates may be simple or complex. The simplest are one-sugar molecules called *monosaccharides*, such as *glucose, fructose,* or *galactose*. White or brown table sugar is *sucrose*, which is a two-sugar molecule, or *disaccharide*, formed from the combination of

Is Fiber a Nutrient?

The cell walls of plants are made of large carbohydrate molecules called *cellulose*. Cellulose and several other large carbohydrate molecules are known collectively as *fiber*. Fiber cannot be digested by human beings (but can be by cows and other ruminant animals), and so it passes through our digestive systems without breaking down the way other foods do. Although not considered a nutrient, the roughage found in fibrous skins and stems of fruits and vegetables is beneficial to our health: Like a broom, fiber sweeps away old food particles as it moves through our digestive systems. And because it cannot be metabolized and absorbed into the bloodstream, it adds few calories to our diets. That is why people who are trying to reduce their weight are advised to eat an abundance of fruits, vegetables, and other foods high in fiber. While providing a full, satisfied feeling, fiber passes right through and adds no fat to the body.

glucose and fructose. Plant foods rich in simple carbohydrates include sugar cane, sugar beet, maple syrup, and fruits.

Complex carbohydrates are long chains of sugar units called *polysaccharides.* Starch, a polysaccharide formed by the connection of glucose molecules, is the primary storage form of carbohydrates in plants. Foods such as bread, rice, oats, pasta, potatoes, cassava, corn, and yams are some examples of the world's most important sources of complex carbohydrates in the human diet. Every culture has its signature starchy food, usually grain or root crops. These crops, such as rice in Asia, provide people not only with a major part of their calories, but also an important sense of cultural identity. (At the beginning of this unit is a picture of Asmaru, who studied cassava, a staple food she knew as a child in Sierra Leone.)

The single source of all carbohydrates in plants is photosynthesis, the conversion of solar energy into chemical-bond energy in plants. When people (or animals) digest foods, the energy being released is what enables us to work. The potential for a food to supply energy is measured in calories. Carbohydrates contain approximately 4 calories per gram. For most healthy people, carbohydrates (preferably complex) should provide about 55 % of total daily calories.

Rice at the Center

Rice is the central grain in the diets of many people of the world, and in these cultures a meal just isn't complete without it.

Carol Pierce Colfer, an anthropologist who studies the forest management practices of the Uma' Jalan Kenyah people of East Kalimantan, Indonesia, tells the story of a dinner she went to with her research collaborator Tamen Uyang, an Uma' Jalan scientist. (From *Beyond Slash and Burn: Building on Indigenous Management of Borneo's Tropical Rain Forests*, by Carol Pierce Colfer [The New York Botanical Garden Press, Bronx, 1997]):

"What do you eat in America? Is it true that you only eat bread?"

"No," I explain, "each meal normally has several things, a meat, a vegetable, and something like rice or bread or noodles."

How many times had I had this conversation? Any Kenyah meal consists of roughly 80% rice, with a small handful of some other item (leaves, meat, noodles, bamboo shoots, ferns). When I tried to explain American diets, the general response was bewilderment. They really couldn't conceive of living without eating rice at every meal.

One day a visiting German team came up the river and invited Tamen Uyang and me for a meal aboard their comparatively posh research vessel. I was ecstatic at the prospect and equally ecstatic at the reality. Steak, baked potatoes, green beans, and other Western goodies were a welcome change. I stuffed myself and noted that Tamen Uyang ate just as much as or more than I did.

On our way home Tamen Uyang asked if we might stop at a small *warung* (a one-table café). A bit surprised, I said sure, what did he want to get? He looked a bit sheepish and said, "A plate of rice. I'm just never quite satisfied if I haven't had any rice."

Fats

Fats, technically known as *lipids*, are the most concentrated sources of food energy. They contain, by weight, more than twice the number of calories (9 per gram) that carbohydrates contain. Although often maligned as nutritional villains, fats perform several important body functions. They supply the body with *essential fatty acids*, the raw materials needed to make hormone-like compounds that help control blood pressure, blood clotting, and inflammation, among other things. They transport fat-soluble vitamins (A, D, E, and K) through the bloodstream, regulate blood cholesterol levels, help keep skin supple and soft, and protect the body from temperature extremes. Fats also store excess calories that the body draws upon when carbohydrate stores are depleted.

Most of the fats in foods are *triglycerides*: three fatty acid molecules bound together by one glycerol molecule. *Fatty acids* are long molecular chains made up of carbon, hydrogen, and oxygen atoms. These chains vary in length and in whether or not every link of the chain has two hydrogen atoms. This variation affects their physical properties in a way that you probably see every day: At room temperature *saturated fats*, which have two hydrogen atoms on every link in its fatty acid chains, are solid, like butter and lard; *unsaturated fats*, which have one or more links with only one hydrogen atom, are in a liquid state at room temperature, such as olive oil and vegetable oil.

Saturated fats are derived chiefly from animal sources such as meat, lard, butter, and other dairy products. Two vegetable oils — coconut oil and palm oil — are also highly saturated. A diet high in saturated fats can result in an increase of the type of blood cholesterol (LDL) that can lead to heart disease and hardening of the arteries.

Unsaturated fats are more common in plants than in animals and include oils derived from olive (*Olea europa*), canola (*Brassica napus*), peanut (*Arachis hypogea*), safflower (*Carthamus tinctorius*), soybean (*Glycine max*), corn (*Zea mays*), and sesame (*Sesamum orientale*). Fish are also an important source of unsaturated fats. The majority of the fats we consume should be unsaturated. Research shows that some unsaturated fats such as olive oil may actually reduce the harmful LDL cholesterol in the blood. The American Heart Association recommends that saturated fats make up no more than 10% of the calories in our diet, and that total fat consumption not exceed 30% of our daily caloric intake.

olive (*Olea europa*)

Protein

Protein makes up 75% of our body tissue (if you don't include water in the tissues). Muscle, skin, bone, nails, some hormones, and all enzymes are composed largely of proteins. Proteins hold an organism together and run it; they repair and construct body parts.

Proteins are made up of individual amino acids linked together in long chains. Shorter strings of amino acids (the result of protein digestion) are called polypeptides. Twenty-two different amino acids have been identified in the human body. Each amino acid group includes carbon, hydrogen, and oxygen; but unlike carbohydrates or fats, each amino acid group also contains nitrogen. These amino acids can combine in a nearly infinite variety of ways. Consider this: A single cell of your body may contain 10,000 different proteins, each one synthesized by linking together a distinct arrangement of amino acids.

Nine amino acids are determined to be *essential* in human nutrition because they cannot be manufactured by the body. They must be supplied in the foods we eat. Proteins that supply all the essential amino acids in the proper amounts are called *complete proteins*. Virtually all animal proteins are complete proteins. On the other hand, plant proteins, except for soybeans, lack one or more essential amino acids. Nonetheless, plant foods can still be excellent sources of protein when eaten in the right combinations.

Traditional diets throughout the world, since the dawn of agriculture 10,000 years ago, have included combinations of plant foods that supply all the essential amino acids.

soybean (*Glycine max*)

When taken together, a grain crop and a legume (plant from the bean family) make a complete protein. Corn tortillas and beans from Mexico; curried lentils and rice from India, rice and tempeh (fermented soybean cake) from Indonesia, and bread and peanut butter from the United States are just a few examples.

Tortillas and Beans Were Meant for Each Other

As agriculture developed independently in different parts of the globe, one finds that legumes (plants in the bean family) were domesticated along with the grains. Legumes are an important source of protein, and are particularly rich in the amino acids lacking in grains. In the Mediterranean, peas and lentils accompanied the domestication of wheat and barley. In the Americas, kidney beans and lima beans were domesticated alongside corn some 7,000 years ago. And in Asia, soybeans seem to have accompanied the cultivation of rice and millet.

So go ahead and enjoy that bean burrito, pita with chickpeas, rice and beans, or peanut butter sandwich knowing that you are eating a nutritious meal made up of complementary proteins. Agriculture evolved from the beginning to serve you such complete and tasty dishes.

Vitamins

Vitamins are complex organic substances required in small amounts that help regulate functions within cells. Vitamins affect a multitude of tasks including the formation of normal blood cells, the proper functioning the heart and nervous system, and the promotion of good vision. For the most part, our bodies do not manufacture these vitamins (except for D in the presence of sunlight, and K), so we need to get them in our food.

Vitamins are grouped into two different classes: fat-soluble, including vitamins A, D, E, and K; and water-soluble vitamins including the B-complex and C. The body can store fat-soluble vitamins for relatively long periods of time, while water-soluble vitamins aren't stored in significant amounts and need to be replenished on a regular basis.

In the following table you will find a short summary of these vitamins, their key functions in the body, and some common sources for each. In the lab exercises on nutrition, you will test your plant for the presence of vitamin C, one of the most important vitamins.

Vitamin	Function/use in humans	Source
A (retinol)	Helps build & maintain healthy teeth, bones, soft tissue, skin, and mucous membranes; promotes vision.	Eggs, meat; also derived from foods high in beta-carotene, such as carrots, sweet potatoes, cantaloupe, broccoli, spinach.
B_1 (thiamin)	Helps the body convert carbohydrates into energy; essential for functioning of heart, muscles, and nervous system.	Soybeans, lean meats (especially pork), fish, peas, pasta, whole grains, (especially wheat germ), cereals, fortified bread.
B_2 (riboflavin)	Helps the body convert carbohydrates into energy; assists in production of red blood cells.	Lean meats, eggs, dairy, legumes, grains, nuts, green leafy vegetables.
B_3 (niacin)	Promotes good functioning of digestive system, skin, and nerves; helps convert food into energy.	Lean meats, poultry, fish, dairy, nuts, eggs.
B_6 (pyridoxine)	Helps maintain normal brain function and red blood cell formation; required for the chemical reactions of proteins; assists in the synthesis of antibodies.	Nuts, legumes, fish, meats, eggs, whole grains, fortified breads and cereals.
B_9 (folate)	Required for red blood cell production, tissue growth, cell function, synthesis	Legumes, citrus fruits, whole grains, dark green leafy vegetables, poultry, pork,

	of DNA; acts as a coenzyme to metabolize proteins; stimulates digestive acid.	shellfish, liver.
B_{12}	Plays important role in metabolism; promotes red blood cell formation and nervous system functioning.	Dairy, eggs, meat, poultry, shellfish.
Biotin	Helps metabolize proteins and carbohydrates.	Eggs, fish, dairy, legumes, whole grains, broccoli, cabbage, lean beef.
Pantothenic acid	Plays a role in the synthesis of hormones and cholesterol.	Eggs, fish, dairy, legumes, whole grains, broccoli, cabbage, lean beef.
C (ascorbic acid)	Promotes healthy teeth and gums; assists the immune system; helps absorb iron and maintain healthy connective tissue.	Citrus fruits, green peppers, broccoli, tomatoes, cantaloupe, strawberries, potatoes.
D	Helps the body absorb calcium, which is essential for healthy teeth and bones.	Dairy, fish, oysters, fortified cereal.
E (tocopherol)	Protects tissue from the effects of oxidation; helps red blood cell production and the absorption of vitamin K.	Wheat germ, nuts, seeds, olives, vegetable oil, spinach and other green leafy vegetables, asparagus.
K	Plays an essential role in blood clotting and the formation of blood-clotting proteins.	Leafy green vegetables, olive oil, soybean oil, liver, green tea.

Minerals

Minerals are inorganic nutrients needed in small quantities to regulate body processes and to help give your body structure. Minerals make up approximately 4% of our body weight but are present in every cell. They participate in such key processes as bone formation, heart function, and production of enzymes. Minerals in our bodies derive originally from the earth's crust. They may have been eroded by ground water or sea water, eventually being taken up by plants to be eaten in foods. Minerals, unlike vitamins, cannot be destroyed by heat.

Our bodies contain about sixty different minerals, although only about twenty-two are considered essential. They fall into two categories: *macrominerals*, which are needed in relatively large quantities, and *trace nutrients*, which are required in very minute amounts. In the table on the following page you will find the seven major nutrients (calcium, chloride, magnesium, phosphorus, potassium, sodium, and sulfur) and two selected trace nutrients (zinc and iron) listed with their functions and sources. In the lab activity, you will discover what percentage of your study plant is composed of minerals.

Filling up with Junk

So-called junk foods are foods that may be high in carbohydrates and fat but do not have many vitamins and minerals. These foods are usually highly processed and contain white flour, refined sugar, and excessive amounts of fat, sodium, and preservatives. Potato chips, cheese curls, french fries, candy, soda pop, doughnuts, cookies — these things may fill you up, but they don't give your body the nutrition it wants. The calories people get from junk food are "empty calories," because they contain little or no nutritional value. Excessive consumption of junk food can contribute to certain diseases such as obesity, diabetes, and heart disease.

The next time you need a "sugar fix," reach for an apple, orange, or peach, or some honey drizzled on a rice cake; or if your stomach is grumbling between meals, you can tide yourself over with peanut butter on either a stalk of celery or a whole-wheat cracker.

Mineral	Function/use in humans	Source
Calcium	Promotes healthy bones and teeth; helps maintain nerve function and blood clotting.	Dairy, tofu, legumes, broccoli, salmon.
Chloride	Helps maintain fluid balance and digestion.	Table salt, soy sauce.
Iron	Promotes red blood cell formation and muscle function; helps the body convert food into energy.	Meat, fish, shellfish, legumes, dried fruits.
Magnesium	Promotes bone health, protein synthesis, enzyme reactions, muscular function, nerve transmission.	Legumes, whole grains, nuts, dark green vegetables, chocolate.
Phosphorus	Helps maintain body pH and DNA/RNA structure.	Dairy, meat, fish, eggs, legumes, whole grains.
Potassium	Promotes muscle function and nerve transmission; helps maintain fluid balance and steady heartbeat.	Fruits, vegetables, dairy, grains, beef, legumes.
Sodium	Helps maintain body pH and fluid retention; helps nerve transmission.	Table salt, soy sauce, most processed foods.
Sulfur	Acts as a component of insulin, biotin, thiamin, some amino acids.	All protein-containing foods.
Zinc	Promotes wound healing, healthy fetal development, sense of taste; plays a role in enzyme function.	Meat, fish, grains, vegetables.

Water

Water, the most abundant substance in the human body, is our most essential nutrient. Without food we can live for few weeks; without water we would perish within a few days. Between 55 % and 75 % of the human body is made up of water, as much as ten to twelve gallons in an adult's body. Two to three quarts must be replaced each day. We can replace some of the water we need by eating food which generally contains more than 85 % water. We should drink an additional six to eight glasses of liquid (such as water, juice, tea, soup, and milk) to make up the balance.

Water gives life to all cell processes and organ functions, forms the basis of all body fluids, and regulates body temperatures. An essential lubricant, a beauty aid to the skin, a cushion for your joints, a means of transporting nutrients — the list of services that water performs is enormous!

Thirst is the body's signal to replenish this nutrient. We feel thirst when we lose just 1 % of our body's water. When we lose 2 % to 5 % of our body weight in water, we feel fatigue, headache, flushed skin, dry mouth, and a reduction in physical performance.

A final note: When assessing the potential benefits of your specific study plant, consider also the role and importance of water. Consider (1) the water that is contained in the plant itself if consumed; and (2) the role of water to prepare and use your plant (for instance, in an extract or tincture). In the lab exercises at the end of this chapter, you will find out what percentage of your plant is composed of water.

Unit 4 Questions for Thought

On a separate piece of paper, answer the following questions as thoroughly as possible.

1. Describe a healthy meal your family might eat, and then answer the following questions:
    a. Which culture does this meal come from?
    b. What are the carbohydrate sources in this meal? List sources of sugars and starches.
    c. What is the fat or oil in this meal?
    d. What is the protein in this meal?
    e. What is the drink (source of water) in this meal?
    f. Is there a source of fiber in this meal? If so, what is it?
    g. If you know which vitamins and minerals are supplied by this meal, list them.

2. Take a look at the three meals described below and answer the questions about them. Each meal includes a cup of orange juice for the beverage. The portions are of average size.

	Meal 1 Hamburger, fries, and ice cream	Meal 2 Chicken, rice, and broccoli	Meal 3 Tossed salad and two slices of pepperoni pizza
Calories	938	510	800
Protein (g)	35	33	25
Total fat (g)	47	5	41
Saturated fat (g)	20	1	15
Unsaturated fat (g)	27	4	26
Total carbohydrate (g)	101	80	82
Sugar (g)	48	25	24
Starch (g)	42	46	48
Fiber (g)	11	9	11
Sodium (mg)	1,015	505	600

a. Which of these meals has the most calories?
b. Which has the most protein?
c. Which would be good for someone trying to lose weight?
d. Which would be best for a diabetic, who needs to avoid carbohydrates and sugar?
e. Which would be best for someone with high blood pressure, who needs to avoid sodium and saturated fats?
f. Which has the most fiber that would help someone with digestive disorders?
g. Which is the most appetizing to you?
h. Which is the healthiest meal overall?

3. List two foods made from plants that are high in each of the following nutrients:
    a. Sugar
    b. Starch
    c. Protein
    d. Oil
    e. Vitamin C
    f. Calcium and magnesium
    g. Iron

4. What is the role of each of the three types of carbohydrates in the body?
    a. Sugar
    b. Starch
    c. Fiber

5. What is the role of protein in the body?

6. What is a complete protein and how can a vegetarian get a complete protein without eating meat?

7. Why should someone who is dieting not completely eliminate fats from the diet?

8. Why is it important to drink 6 to 8 glasses of liquids every day?

9. Why are some foods called "junk food," and why are these foods so bad for you?

10. Take a look at the food pyramid below and answer the following questions:
    a. About how many servings of foods from plant products (including fruit, vegetables, nuts, grains, beans, and sweets) does it recommend that you eat daily?
    b. About how many servings of foods from animal products (including meat, fish, eggs, dairy products, and seafood) does it recommend that you eat daily?

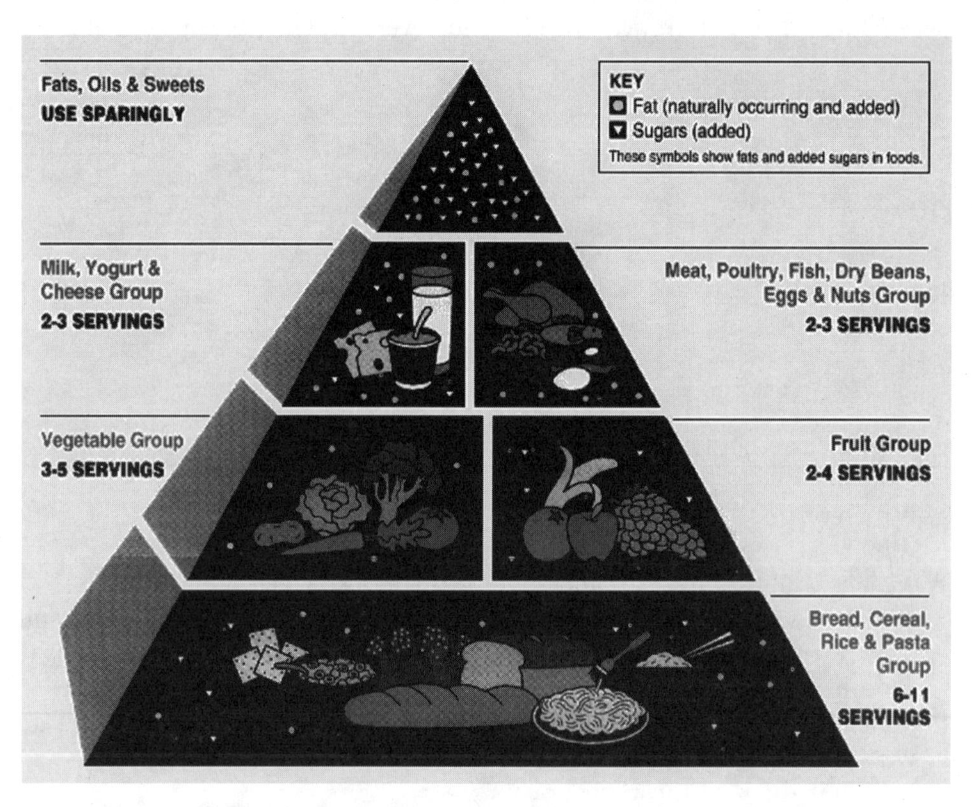

illustration courtesy of U.S. Department of Agriculture and U.S. Department of Health and Human Services

Unit 4 Activities

Field Exploration: Comparing Food Labels

Name: ______________________ Date: ______________________

You can learn a lot about the nutritional benefits of foods by examining their labels. In this activity you will compare the nutrition information from three different products. Compare three foods that you might choose for a meal or snack. For example, compare three different snack foods (such as cookies, crackers, and chips), three types of breakfast cereals, or three dairy products, or soups, or pasta dishes. You may want to compare very different foods, such as a highly refined food and a whole-grain food, or breakfast foods and dinner dishes. Compare any three foods you like, but have a reason for your comparison.

	Product #1	Product #2	Product #3
Product name:	________	________	________
Ingredients derived from:			
plants:	________	________	________
animal products:	________	________	________
Other ingredients:	________	________	________
Serving size:	________	________	________
Calories:	________	________	________
Calories from fat:	________	________	________
Total fat:	________	________	________
Saturated fat:	________	________	________
Sodium:	________	________	________
Potassium:	________	________	________
Total carbohydrates:	________	________	________
Dietary fiber:	________	________	________
Sugars:	________	________	________
Other carbohydrates:	________	________	________
Protein:	________	________	________
Vitamins present in amounts of at least 5 % of the required daily allowance (RDA):	________	________	________
Minerals present in amounts of at least 5 % of the RDA:	________	________	________

Answer the following questions about these products:

1. Why did you choose these three products for comparison with one another?

____

2. Which product has the most fat? ______________________
3. Which product has the most saturated fat? ______________________
4. Which product has the most calories? ______________________

5. Which product has the most sugar? ______________________________

6. Which product has the most protein? ______________________________

7. Which product contains the most fiber? ______________________________

8. Which product has the most vitamins? ______________________________

9. Is this a vitamin-fortified food? (If so, vitamins will be listed by name as part of the ingredients.) ______________________________

10. Which product has the most minerals? ______________________________

11. Which product do you consider to provide the most nutritional benefits and why?

____

____

12. Which product do you consider to be of the least nutritional value and why?

____

____

Laboratory Activities: Testing Your Plant for Selected Nutrients

In the following experiments you will discover if your study plant contains any of the major groups of nutrients discussed in this chapter. For purposes of comparison, you will compare your plant extract (sample A) to a negative control without the nutrient (sample B) and a positive control that has the nutrient (sample C).

General materials needed

Section of fresh plant, plant extract, or dried plant (for instructions on making a plant extract or dried plant, see Unit 3 Activities)

Nutrition testing kit (see Appendix A for sources) or materials listed under each experiment

(Note: For the mineral and water tests that follow, a section of fresh plant material is needed. The water test should be conducted first, so that after the plant material is dried out from the water test, it can be used to conduct the mineral test.)

A. Does Your Plant Have Starch?

Name: ______________________________ Date: ________________________

In this experiment you will find out whether or not your plant has starch.

Materials needed

Three test tubes
Your plant extract
Small amount of warm water in which a piece of bread has been soaking
Beaker full of pure (preferably distilled) water
Wax pencil
Small ruler labeled with centimeters
Five medicine droppers
Iodine or Lugol's solution

Directions

1. With the wax pencil, mark each test tube 2 cm from the bottom. Label them A, B, and C.

2. Add 2 cm of water to each tube. To Tube A add 10 drops of your plant extract. Add 10 drops of water to Tube B and 10 drops of the water that held the bread to Tube C. Use a different medicine dropper for each solution so they do not get mixed together.

3. Using the last medicine dropper, add 4 drops of the iodine solution to each tube. Shake the tubes gently to mix them. By comparing the tubes B (without starch) and C (with starch) to the one with your plant extract, you will be able to tell whether or not your plant has starch. A color change to blue-black indicates the presence of starch.

Analysis

1. Plant being studied: ______________________________
2. What was in test tube A? ______________________________
3. What was in test tube B? ______________________________
4. What was in test tube C? ______________________________
5. What color did tube A turn after adding iodine? ____________________
6. What color did tube B turn after adding iodine? ____________________
7. What color did tube C turn after adding iodine? ____________________
8. Does your plant extract have starch? How can you tell? ________________

____

B. Does Your Plant Have Sugar?

Name: ______________________________ Date: ________________________

In this experiment you will find out whether or not your plant has sugar.

Materials needed

Copper (II) hydroxide in an aqueous medium with sodium citrate or Benedict's solution
Three test tubes
Your plant extract
Small amount of warm water saturated with sugar
Beaker full of pure (preferably distilled) water
Bath of hot water
Wax pencil
Small ruler labeled with centimeters
Five medicine droppers

Directions

1. With the wax pencil, mark each test tube at 2 cm from the bottom. Label them A, B, and C.

2. Add 2 cm of water to each tube. To Tube A add 10 drops of your plant extract. Add 10 drops of water to Tube B and 10 drops of sugar water to Tube C. Use a different medicine dropper for each solution so they do not get mixed together.

3. Using the last medicine dropper, add 20 drops of Copper (II) hydroxide/sodium citrate solution (or Benedict's solution) to each tube. Shake the tubes gently to mix them.

4. Put all three test tubes into the hot-water bath for about five minutes. By comparing the tubes B (without sugar) and C (with sugar) to the one with your plant extract, you will be able to tell whether or not your plant has sugar. A color change from blue to yellow, orange, or red indicates the presence of sugar.

Analysis

1. Plant being studied: ______________________________
2. What was in test tube A? ______________________________
3. What was in test tube B? ______________________________
4. What was in test tube C? ______________________________
5. What color did tube A turn after being heated? ____________________
6. What color did tube B turn after being heated? ____________________
7. What color did tube C turn after being heated? ____________________
8. Does your plant extract have sugar? How can you tell? ____________________

____

C. Does Your Plant Have Oil?

Name: ______________________ Date: ______________________

In this experiment you will find out whether or not your plant has oil.

Materials needed

A paper bag cut into three strips
Your plant extract
Small amount of warm water
Small amount of corn oil or other common oil
Pencil
Three medicine droppers

Directions

1. With the pencil, label the paper strips A, B, and C.

2. Put a drop of your plant extract onto Strip A. Put a drop of water onto Strip B and a drop of oil onto Strip C. Use a different medicine dropper for each solution so they do not get mixed together. Let the paper dry for a few minutes.

3. Hold the strips of paper up to the light. The one(s) with oil will be slightly translucent and shiny. The one(s) with no oil will have dried out.

4. Rub your finger along the part of each strip where the drop was placed. Oils feel greasy and smoother than water. You should be able to feel the water.

Analysis

1. Plant being studied: ______________________
2. What was on paper strip A? ______________________
3. What was on paper strip B? ______________________
4. What was on paper strip C? ______________________
5. What did you see when you held strip A to the light?______________________
6. What did you see when you held strip B to the light? ______________________
7. What did you see when you held strip C to the light? ______________________
8. How did strip A feel when you rubbed your finger on it? ______________________
9. How did strip B feel when you rubbed your finger on it? ______________________
10. How did strip C feel when you rubbed your finger on it? ______________________
11. Does your plant extract have oil? How can you tell? ______________________

____

D. Does Your Plant Have Protein?

Name: ______________________________ Date: ________________________

When sodium hydroxide and copper sulfate are mixed in the presence of protein, a pink or purple precipitate forms. A negative test produces a blue precipitate with the liquid remaining clear.

In this experiment you will find out whether or not your plant has protein.

Materials needed

Three test tubes
Your plant extract
Small amount of warm water mixed with an equal amount of milk
Beaker full of pure (preferably distilled) water
Wax pencil
Small ruler labeled with centimeters
Six medicine droppers
Sodium hydroxide solution
Copper sulfate solution

Directions

1. With the wax pencil, mark each test tube 1.5 cm from the bottom. Label them A, B, and C.

2. Add 1.5 cm of water to each tube. To Tube A add 10 drops of your plant extract. Add 10 drops of water to Tube B and 10 drops of the water and milk mixture to Tube C. Use a different medicine dropper for each solution so they do not get mixed together.

3. Using the last medicine droppers, add 15 drops of sodium hydroxide and 15 drops of copper sulfate to each tube. The solution should start out blue. By comparing tubes B (without protein) and C (with protein) to A (with your plant extract), you will be able to tell whether or not your plant has protein. If the solution turns from blue to pink, protein is present; if the solution remains blue, no protein is present.

Analysis

1. Plant being studied: ______________________________
2. What was in test tube A? ______________________________
3. What was in test tube B? ______________________________
4. What was in test tube C? ______________________________
5. What color did tube A turn after adding the chemicals? ______________
6. What color did tube B turn after adding the chemicals? ______________
7. What color did tube C turn after adding the chemicals? ______________
8. Does your plant extract have protein? How can you tell? ______________

____

E. Does Your Plant Have Vitamin C?

Name: ______________________________ Date: ________________________

In this experiment you will find out whether or not your plant has vitamin C.

Materials needed

Three test tubes
Your plant extract
Small amount of freshly squeezed lemon juice
Beaker full of pure (preferably distilled) water
Wax pencil
Four medicine droppers
Vitamin C testing solution of 2,6-dichloroindophenol, sodium salt

Directions

1. With the wax pencil, label the test tubes A, B, and C.

2. Add 15 drops of the vitamin C testing solution to each tube.

3. With a new medicine dropper, add your plant extract to tube A one drop at a time. Shake the tube gently after each drop until the solution becomes clear. Don't pay attention to the pink color. The faster it gets clear, the more vitamin C it contains. If it does not have vitamin C it will remain blue. Since vitamin C is a reducing agent, the indophenol compound, which is blue in solution, will become colorless as it is bleached by vitamin C.

4. Using a new medicine dropper, add water one drop at a time to tube B and follow the same procedure as you did with tube A.

5. Using a new medicine dropper, add lemon juice one drop at a time to tube C and use the same procedure as with the other tubes. By comparing the tubes B (with water) and C (with lemon juice, which has vitamin C) to A (with your plant extract), you will be able to tell whether or not your plant contains vitamin C.

Analysis

1. Plant being studied: ____

2. What was in the test tubes at first? ___________________________________

3. How many drops did you have to add to each tube until the solution turned clear?

   Drops of plant extract added to tube A: ________

   Drops of water added to tube B: ________

   Drops of lemon juice added to tube C: ________

4. Does your plant extract have vitamin C? How can you tell? _____________________

____

F. What Percentage of Your Plant Is Made Up of Water?

Name: ______________________ Date: ______________________

In this experiment you will find out what percentage of your plant is made up of water. You will need freshly cut plant material to conduct this experiment, since dried material obviously will have lost much of its water.

Materials needed

Freshly cut material from your plant
Sensitive scale
Heat-resistant bowl
Oven

Directions

1. Weigh the bowl, then fill it with your fresh plant material and weigh it again. Record these weights in the Analysis section, below.

2. Heat the plant in the oven at about 200°F for several hours, or until the plant is totally dry and crispy.

3. Reweigh the bowl with the plant material inside. Record these weights in the Analysis section, below.

Analysis

1. Plant being studied: ______________________

2. How much did the container alone weigh? __________

3. How much did the container and plant weigh together before heating? __________

4. How much did the plant weigh before heating? (Subtract container weight in #2 from total weight in #3.) __________

5. How much did the container and the dried plant together weigh? __________

6. How much did the dried plant alone weigh? (Subtract container weight in #2 from total weight in #5.) __________

7. How much water weight was lost? (Subtract dried plant weight in #6 from fresh plant weight in #4.) __________

8. Determine the percentage of water in the plant by dividing the final weight of the plant (from #6) by the original weight of the plant (from #4). (For example, if the original weight was 2 grams and the final weight was 1 gram, then the percentage of minerals would be 1 ÷ 2, or 0.5; multiply by 100 to find the percent, and you get 50%.)

__________ ÷ __________ = __________ × 100 = ______%

dried weight fresh weight

9. What environmental factors can you think of that might effect the amount of water in the plant? ______________________

____

____

G. What Percentage of Your Plant Is Made Up of Minerals?

Name: ______________________________ Date: ________________________

In this experiment you will see if your plant has minerals. It will not identify the specific minerals contained in the plant, since testing for this requires more sophisticated equipment. When burned, plant materials release carbon dioxide, nitrogen, and water vapor. All the other minerals will be left behind as ash.

Materials needed

Your plant in a dried form
Spoon or crucible
Tongs
Candle or Bunsen burner
A match
Sensitive scale

Directions

1. Weigh the spoon or crucible, then fill it with your dried plant material and weigh it again. Record these weights in the Analysis section, below.

2. Light the candle or Bunsen burner.

3. Hold the spoon or crucible filled with the plant over the flame with the tongs so that the plant will burn. Let the plant burn until there are only ashes remaining. This is the mineral component of the plant.

4. Reweigh the spoon or crucible with the plant ashes still on it. Record these weights in the Analysis section, below.

Analysis

1. Plant being studied: ____
2. How much did the container alone weigh? ___________
3. How much did the container and plant weigh together before burning? ___________
4. How much did the plant weigh before burning? (Subtract the container weight in #2 from the total weight in #3.) ___________
5. How much did both the container and the plant ash weigh? ___________
6. How much did the plant ash weigh? (Subtract the container weight in #2 from the total weight in #5.) ___________
7. How much plant weight was lost due to the burning? ___________
8. Determine the percentage of minerals in the plant by dividing the final weight of the plant ash (from #6) by the original weight of the plant (from #4). (For example, if the original weight was 2 grams and the final weight was 1 gram, then the percentage of minerals would be 1 ÷ 2, or 0.5; multiply by 100 to find the percent, and you get 50%.)

__________ ÷ __________ = __________ × 100 = ______%

ash weight   dried weight

Summary of the Nutrition Experiments

Name: ________________________ Date: ________________________

1. Plant being studied: ________________________
2. Culture(s) in which the plant is used: ________________________
3. Which of the following did your plant have?
   - starch ________
   - sugar ________
   - oil ________
   - protein ________
   - vitamin C ________
4. What was the percent weight of the following?
   - minerals ________
   - water ________

If you did multiple nutrient testing trials and/or different people in a class conducted the nutrient tests on different plants, you can fill in the following chart to record all the data that was collected, including the results of the positive and negative controls.

Nutrient test results for various plants

Plant name	Starch	Sugar	Fat	Protein	Vit. C	Minerals (%)	Water (%)

Note: If you would like to share these experimental results with other interested students, teachers, and scientists on the World Wide Web, you may email the author at the following address to receive instructions on how to do so: gdpaye@hotmail.com.

Unit 5
Testing the Medicinal Properties of Your Plant

Plants as Medicine

Throughout most of human history — up until the last half of the twentieth century — plants have been the primary source of drugs used to treat illness. Even today, in developing countries where pharmaceutical drugs are often scarce or beyond the means of all but the wealthiest individuals, about 80% of people still rely on traditional healing for their health care needs, and the majority of this kind of treatment consists of plant-based medicines.

A glance at the shelves of any pharmacy in the United States and Europe will reveal that there, too, the connection between plants and medicine is a strong one: Approximately one in four prescription drugs contains ingredients derived from plants. In fact, Western medicine has received some 7,000 different medicinal compounds from plants including heart drugs, analgesics, antibiotics, anticancer compounds, anti-inflammatory medicines, oral contraceptives, hormones, laxatives, and numerous other drugs. Many of these plant-based drugs were discovered by first studying their uses in traditional folk medicine.

The discovery of new drugs based on ethnobotanical leads has great potential. While an estimated 35,000 to 70,000 species of plants have been used by various peoples in the world for medicine, only about 5,000 have been tested for their pharmaceutical potential in laboratories. This unit will help you understand how such laboratory research is conducted and will allow you to begin participating in this exciting process.

In this unit you will learn to conduct experiments that test for some possible medicinal properties of your plant. **Important: Never use plants as medicine without first consulting a licensed medical practitioner!** In the lab you will examine the effects of an extract from your plant on a variety of microorganisms grown in culture. If your plant is able to kill or inhibit the growth of a particular type of microbe, it might also succeed in treating diseases caused by similar organisms. Prior to the lab work, though, we will take a brief look at the types of organisms that cause infectious disease, and the ways scientists study the potential effectiveness of new drugs that might combat them.

Keep in mind that the process of developing a new pharmaceutical agent usually requires ten to fifteen years of research, often at a cost of hundreds of millions of dollars. We can only begin to learn about the medicinal properties of plants within the confines of the curriculum. However, we can learn a lot about how medical research is conducted.

Infectious Disease

Infectious diseases such as chicken pox, measles, typhoid, malaria, AIDS, strep throat, and the common cold are caused by microscopic pathogens that invade the body. These life forms are too small to be seen without the aid of a microscope and are thus referred to as *microorganisms.* Common disease-causing pathogens include bacteria, viruses, and microscopic fungi and protozoa.

Bacteria and Bacterial Disease

Bacteria are single-celled microorganisms that live in the air, on land, and in water. Bacteria have a simpler cell structure than other living organisms; they do not have membrane-bound cellular organelles, and they lack an organized nucleus. They reproduce by dividing in half. Because of these unique characteristics that distinguish bacteria from other living things, they are classified in a separate kingdom, the kingdom Monera.

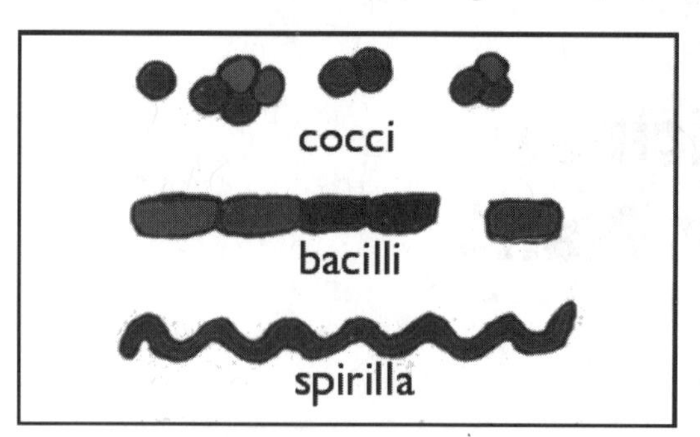

How do scientists and physicians identify different types of bacteria? For example, how can we distinguish the species that causes pneumonia from the one that turns milk into yogurt? Bacteria vary in cell shape, organization, and cell wall structure, so they can be classified by their variation in form and the way the cells adhere to one another. Straight, rod-shaped cells are called *bacilli*, spherical-shaped cells are called *cocci*, and long spiral cells are classified as *spirilla*. Cocci may occur in pairs, clusters, or chains after dividing. Bacilli are more likely than cocci to occur as isolated individuals.

Another key to identification is the ability of a bacterial cell wall to retain a purple color when stained with crystal violet or similar dye. *Gram-positive* bacteria have a thick cell wall made of disaccharides and amino acids that retains the dye. *Gram-negative* bacteria have a cell wall surrounded with an extra layer of lipopolysaccharides that do not absorb the stain; they turn pink instead. Gram-negative and -positive bacteria respond differently to antibiotics. Gram-negative bacteria are often protected from antibiotics such as penicillin or actinomycin by their additional cell wall layer.

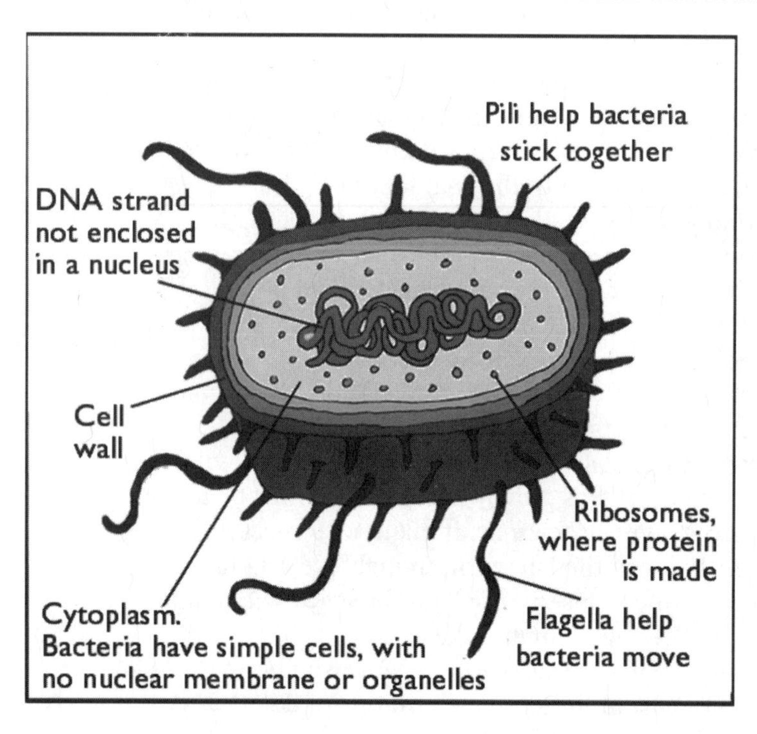

Not all bacteria are harmful. These extremely abundant organisms participate in a multitude of ecological services. Together with fungi, they share the role of decomposers in ecosystems throughout the world, helping to recycle organic matter and nutrients. Bacteria are essential to the nitrogen cycle, and they also assist the cycling of sulfur and carbon. Some bacteria are photosynthetic and make their own food using the sun's energy and chlorophyll, producing oxygen and creating food for other organisms. Chemosynthetic bacteria live deep in the earth or on the ocean floor and derive their energy from inorganic substances. Some types of bacteria live in the human body and aid in food digestion; others turn milk to yogurt, and apple cider to vinegar.

goldenseal (*Hydrastis canadensis*)

Many bacteria are important pathogens. Some of the most devastating plant diseases that plague agricultural crops, such as blights, soft rots, and wilts, are caused by bacteria. Other bacteria cause disease in animals, including people, when they produce toxins that interfere with essential physiological processes. These bacterial pathogens are transmitted in a variety of ways.

Human diseases caused by airborne bacteria include those that cause pneumonia, whooping cough, and tuberculosis. Bacterial pathogens spread through unclean food and water can cause cholera, typhoid fever, and common forms of traveler's diarrhea. Bacteria involved in food spoilage can result in food poisoning, such as botulism and *Salmonella* poisoning. Sexually transmitted bacterial diseases include gonorrhea and syphilis. Insect-borne bacterial diseases carried by fleas, lice, and ticks are responsible for such illnesses as the bubonic plague, typhus, and Lyme disease.

In the twentieth century we made great strides in the control of bacterial disease. Clean water and sanitation, immunization programs, and antibiotics have helped increase the life expectancy of Americans by almost thirty years since 1900. However, we must not become too complacent about these advances. Microbes are able to mutate and adapt to changing environmental conditions. In general, the increasing problem of antibiotic-resistant microorganisms arises frequently as a subject of concern among medical professionals. Of particular concern in the United States is the appearance of new strains of tuberculosis that are resistant to antibiotics.

The Bulb that Wards Off More than Vampires

Research shows that garlic has remarkable antibiotic properties. Garlic was used in World War I to treat wounds, and prior to the development of antibiotics it was used to treat tuberculosis and typhoid. In addition, garlic also provides many heart-healthy benefits. Garlic reduces blood (lipid) fat levels and lowers blood pressure. It also may prevent plaque build-up in the arteries.

In the future, new antibiotics may be developed from plants that have antibacterial properties. Garlic (*Allium sativum*) is an example of a plant that exhibits antibiotic action. This plant has been revered since ancient times for its healing powers. Other plants that have shown promise due to their antibacterial properties are tea-tree (*Melaleuca alternifolia*), neem (*Azadirachta indica*), goldenseal (*Hydrastis canadensis*), onion (*Allium cepa*), turmeric (*Curcuma longa*), and lemon (*Citrus limon*).

Viruses and Viral Diseases

The spread of HIV, the human immunodeficiency virus that causes AIDS (acquired immunodeficiency syndrome), has made viruses a common topic in the media, from news reports to round-table discussions. Viruses have unique characteristics that make them particularly effective, and often deadly, pathogens to both plants and animals.

Viruses are much smaller than bacteria. They consist of only a coat of protein that surrounds an inner core of DNA or RNA. Viruses replicate by inserting their genetic material into the host cell and seizing control over its genetic machinery. The host cell is then forced to produce more viruses, effectively becoming a virus factory. Because viruses cannot reproduce independently without invading a host cell, they are not considered living things.

Human diseases caused by viruses spread by a variety of different means. The pathogens that cause measles, mumps, rubella, chicken pox, and influenza may be transmitted on air currents and by contact. Polio and infectious hepatitis spread through contaminated food or water. Herpes and HIV pathogens are most commonly, but not solely, transmitted though sexual contact. The fatal tropical disease, yellow fever, is spread by mosquitoes.

Influenza, or the flu, is one of the most common recurring infectious diseases that surfaces periodically in global epidemics. The influenza epidemic of 1918–1919 spread rapidly around the globe in three waves and took the lives of more than 20 million people in one year. In the winter of 1968–1969, more than 50 million cases of Hong Kong flu were recorded in the United States; it caused 70,000 deaths. Viruses responsible for the flu, like other microorganisms, continually undergo mutation. These mutations intermittently produce new strains of influenza that spread rapidly and threaten global health.

While antibiotics have made great strides in curing bacterial diseases, fewer medications exist to treat viral diseases. Immunization campaigns provide protection against the spread of polio, measles, mumps, and rubella where they are administered. Due to the most successful immunization campaign in history, there have been no reported cases of smallpox since 1977. However, many viral diseases, including AIDS and the newly discovered *Ebola* and *Hanta* viruses, have no known cure.

Ethnobotanical research may provide one avenue for discovering new medicines that treat viral diseases. In 1984, ethnobotanist Paul Cox found that traditional healers in Samoa were using the inner bark of the *mamala* tree (*Homalanthus nutans*, in the Euphorbiaceae family) to treat what appeared to be acute hepatitis. Follow-up laboratory research on this plant at the National Cancer Institute showed that it contains an active ingredient known as *prostratin* that, *in vitro*, protects healthy cells from infection by the HIV-1 virus. The process that links such lab analyses to the actual development of an HIV drug candidate are long and involved. The important first step, however, originated from careful ethnobotanical research on traditional plant use.

Protozoans and Diseases Caused by Protozoans

Worldwide, more people have died from disease caused by protozoans than have died in war. Protozoans share a phylum, the Protozoa (within the kingdom Protista), with other diverse unicellullar and multicellular organisms. Protozoans have more-complex cells than bacteria and contain a nucleus and membrane-bound organelles.

Important members of the kingdom Protista include the photosynthetic phytoplankton that form the basis of the food chain in aquatic systems. However, it is among the non-photosynthetic protists that we find disease-causing pathogens for plants, animals, and people. For example, the organism that caused the Irish Potato Famine discussed in Unit 1 is a protist called *Phytophthora infestans*, which results in late blight in potatoes.

There are more than 50,000 species of protozoa. They live in water and water-based liquids and move around using small tails or cilia. Malaria, which kills more than 1 million people each year, is one of the most well-known and deadly diseases caused by protozoa. Malaria is transmitted by the female *Anopheles* mosquito. She spreads the pathogen *Plasmodium* when she punctures people's skin to draw blood. The *Plasmodium* parasites reproduce in the liver and red blood cells. Symptoms of this illness include cycles of chills and high fevers followed by excessive sweating and delirium. Victims often die.

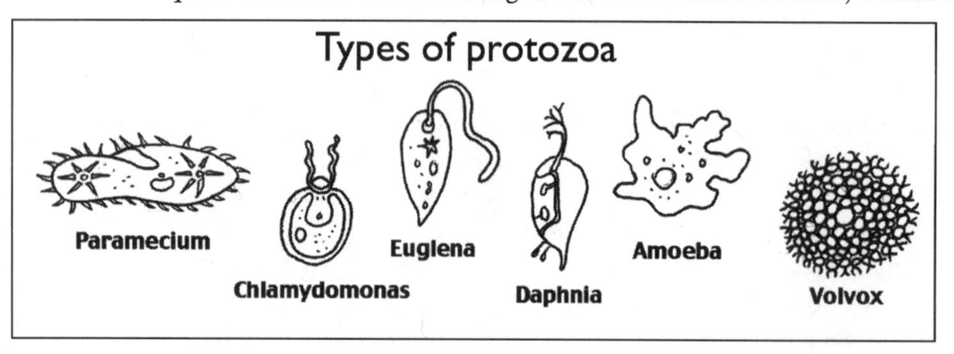

Between the development of the pesticide DDT in the 1950s to kill mosquitoes and the use of quinine-based medicines to treat malaria, this deadly illness was brought under control for many years. However, in many malarial areas we now find mosquitoes resistant to DDT and other pesticides. Additionally, strains of malaria have also appeared that are resistant to conventional drug treatment.

Many other water-borne illnesses are caused by protozoa. *Giardia* causes severe stomach cramping, diarrhea, and nausea. *Giardia* is dreaded by backpackers in the United States who now find many of the wilderness streams contaminated with this organism. Amoebic dysentery is a painful intestinal infection commonly found in developing countries with poor sanitation. Sleeping sickness is a fatal illness in Africa that affects both people and animals. The blood parasite *Trypanosoma* spreads sleeping sickness by tsetse flies. Schistosomiasis is another protozoan illness spread by snails in contaminated water in some tropical countries.

As already noted, quinine from the cinchona tree (*Cinchona pubescens*) has been an important plant-based antimalarial drug. Chinese wormwood (*Artemisia annua*) is said to treat malaria where conventional drug treatments are failing.

Fungi and Fungal Diseases

Molds, mildews, yeast, and mushrooms are all fungi. These remarkable organisms are neither plants nor animals; they differ so greatly from other living things that they form their own distinct kingdom, the Fungi. More than 100,000 species of fungi have been described, and an estimated 1.4 million more await discovery. Although they are often

A Healing Fungus

There are fungi that cause disease in humans, animals, and plants; there are fungi that help plants by breaking down debris into nutrients in the soil. Then there's the fungus that turned out to be a "wonder drug."

Discovered and developed in the 1970s, cyclosporin is a bioactive compound that was isolated from *Tolypocladium inflatum*, a microscopic fungus found in soil. Researchers discovered that cyclosporin helped suppress the immune system. This might not sound like a positive thing, but it was an exciting development for patients who receive organ transplants. Without cyclosporine (the drug made from cyclosporin) or other immunosuppresive drugs, frequently these patients' bodies reject the new organ because the immune system behaves as though the new organ is a foreign object that the body must be protected against. There were immunosuppresants for transplant patients before the development of cyclosporine, and though they are still in use, many of them have more severe side effects than cyclosporine has.

grouped together with plants, fungi do not make their own food as plants do. Some obtain food by feeding on dead organic material (*saprophytes*), and others obtain it by feeding on living matter (*parasites*). Unlike plants, fungi have cell walls made of chitin, the same material that makes up the hard exoskeletons of insects and crustaceans.

Most fungi are multicellular terrestrial organisms that consist of a great many thread-like filaments called *hyphae*. A mass of hyphae is called a *mycelium*. Within 24 hours one fungus may produce one kilometer of mycelium. They reproduce by spores which are spread by wind, water, insects, and animals. With such rapid growth and invasive, thread-like form, fungi are in very intimate contact with their environment, uniquely suiting them for the ecological roles they play. Fungi, together with bacteria, serve as decomposers in the biosphere. Some fungi, the *mycorrhizae*, form symbiotic relationships with plant roots and aid in plant nutrition.

Armed with enzymes that break down organic products, fungi can be destructive as well as helpful. Their success at rotting dead wood also allows them to succeed at rotting fence posts, wooden house foundations, cloth, paper and books, especially in damp environments. Molds can rot bread, fruits, vegetables, and meats. Some molds produce toxins, such as the dangerous aflatoxin that is found in moldy peanuts. It is little surprise that fungi, including rusts, powdery mildews, and damping-off, are the main cause of disease in agricultural crops.

tea tree (*Melaleuca alternifolia*)

Human diseases caused by fungi include thrush, ringworm, athlete's foot, and vaginal yeast infection. These infections cause itching discomfort and sometimes cause blemishes. People with compromised or stressed immune systems are more prone to fungal infections.

In the process of evolution, plants have developed phytochemicals that fight off fungal diseases. Some of these plant compounds may also help in the treatment of human disease caused by fungi. Some common plants that have exhibited an ability to inhibit fungal growth include thyme (*Thymus vulgaris*), garlic (*Allium sativum*), tea tree (*Melaleuca alternifolia*), and lapacho (*Tabebuia* species). Other plants used in traditional treatment of fungal infections, such as echinacea (*Echinacea* species), help boost the immune system so the body is better able to resist infections on its own.

Noninfectious Disease

The illnesses mentioned thus far are infectious diseases caused by pathogens that can be spread from one person to another. Many illnesses are not caused by pathogens but rather by behavioral, genetic, or environmental factors, or a combination of all three. Noninfectious diseases include cancer, heart disease, diabetes, emphysema, arthritis, mental illness, and many others. A thorough discussion of these illnesses is beyond the scope of this book. While no plant medicine has been identified as a cure-all for any of these noninfectious diseases, many plant medicines have been shown to stall the onset or reduce the severity of symptoms of some of these illnesses.

Many noninfectious diseases prevalent in the United States are caused by behavioral factors such as a poor diet, overeating or excessive consumption of fats, smoking, and the abuse of drugs or alcohol. Other health problems such as asthma or respiratory allergies may be triggered by environmental factors such as the increasing contamination of our air, water, and food by pollutants and pesticides. While studying the potential of plant medicines to promote health, it is important to keep in mind the larger picture of our behavioral choices and the overall environment we live in.

In the following chart you will find a few examples of noninfectious diseases and the plant-derived compounds used to treat them. All of these compounds are already components of commercial pharmaceutical drugs.

Illness	Description	Drug	Plant species [family]
Pediatric leukemia	A cancer that causes an excess of white blood cells	Vincristine	*Catharanthus roseus* [Apocynaceae]
Hypertension	High blood pressure	Reserpine	*Rauvolfia serpentina* [Apocynaceae]
Glaucoma	Excess pressure in eye which can lead to blindness	Physostigmine	*Physostigma venenosum* [Fabaceae]
Congestive heart failure	Inadequate pumping action of the heart; atrial fibrillation	Digitoxin and digoxin	*Digitalis purpurea* [Scrophulariaceae]
Hodgkin's disease	Progressive enlargement of lymph nodes and spleen	Vinblastine	*Catharanthus roseus* [Apocynaceae]
Gout	Pain and swelling in feet due to build-up of uric acid	Colchicine	*Colchium autumnale* [Liliaceae]

How Scientists Analyze New Drugs in the Lab

Scientists first test and analyze plant compounds in the lab for their potential as medicines using many different methods. The objective of these tests are basically twofold. First, we want to find out if the compound being tested (such as a plant extract) demonstrates any potential *bioactivity*, that is, its effect on a living organism. For example, if the compound kills bacteria in culture, it demonstrates bioactivity. Second, we want to identify the different components that make up the mixture or compound being tested; in particular, a scientist would try to identify the specific type of molecule responsible for a compound's effectiveness.

A simple but clear example is that of garlic. In lab tests, extracts of garlic have been shown to kill bacteria in culture. Hence, garlic contains *bioactive* compounds. Further testing has shown that the compound responsible for garlic's effectiveness is allicin, the chemical responsible for garlic's powerful smell.

What's a Phytochemical?

Phytochemicals are the chemical compounds found in plants. Some are parts of vitamins and some are parts of minerals, but many phytochemicals have no known nutritional value, and these are the ones that many scientists are focusing on. These bioactive chemical compounds are beginning to be understood in terms of the health benefits they may offer. Scientists are now beginning to discover how these different compounds act at the cellular level and how they may help prevent disease. Some examples of different families of important phytochemicals found in common foods are listed below:

Phytochemical family	Major food source
Allyl sulfides	Onions, garlic, leeks, chives
Indoles	Broccoli, cabbage, kale and cauliflower
Isoflavones	Soybeans and soyfoods
Phenolic acids	Whole grains, seeds, tomatoes, carrots, citrus
Polyphenols	Green tea, grapes
Saponins	Beans and other legumes
Terpenes	Cherries, citrus peel

In the following section we describe how scientists test compounds for bioactivity on microorganisms grown in culture. We also describe three main techniques — chromatography, electrophoresis, and spectroscopy — used to separate out the different components of a mixture. Each of these tests yield useful but limited information. They form small but important steps in the larger effort to discover and analyze new medicines.

Microorganisms in Culture

When provided with an adequate food source and suitable environmental conditions, bacteria, fungi, and protozoans can be grown in culture. By growing specific microorganisms in a container outside their normal host or habitat, they can be studied and subjected to different tests.

Agar, a jelly-like substance extracted from the seaweed shown at right, is the most common medium in which microbes are grown and studied. Agar is heated, mixed with nutrients, and poured into sterile petri dishes or test tubes. When cool, the agar is inoculated with the type of microbe to be cultured. As the desired colonies of bacteria or fungi grow, they can be subjected to different tests (such as exposure to plant extracts) to see how they respond.

agar (*Gelidium amansii*)

Growing a microorganism in culture is like putting a fish in an aquarium. It will be easy to study and will yield some useful information, but many details about how it interacts in its normal environment will remain unknown.

A final note: It is not possible to examine viruses in culture because they cannot reproduce outside their hosts. For this reason, it is more difficult to study viruses than other microorganisms.

The Dirty Dish that Changed the Medical World

In 1929 a scientist named Alexander Fleming was studying bacteria that he had cultured in petri dishes. One of the samples was accidentally contaminated with mold. Upon closer examination, he noticed that the bacteria appeared to be growing throughout the dish except near the blue-green fungus. He hypothesized that the mold must be producing a chemical that suppressed the growth of bacteria, and he went on to discover that this mold that discouraged bacterial growth in a petri dish was also capable of warding off bacterial infections in the human body. That mold was pennicillin, one of the most important antibiotics of the twentieth century. And to think, it all started with a dirty dish!

Chromatography

Plants are complex and contain a myriad of different chemical compounds. In testing for potential drugs, the different compounds must be separated out so that each one can be studied in greater detail. One method commonly used to separate substances in a mixture is chromatography.

In chromatography, the components separate out as they flow across a medium or absorbent. In paper chromatography, the absorbent is paper. In thin layer or gel filtration chromatography, the absorbent is a film of silica gel. Separation occurs because the components in a mixture differ in their attraction to the absorbent and thus they flow across it at distinct rates. Molecules separate out based on their particular size.

Electrophoresis

Electrophoresis is another technique commonly used to separate, identify, and analyze chemical compounds in plants. In this process, molecules are passed through a stationary gel that is exposed to an electric field. The molecules separate and migrate through the electric field. How far the different molecules travel and where they settle in the gel depends on the size and charge of the molecule. By observing the types of behaviors these molecules exhibit, scientists can tell what kind of molecules they are.

Electrophoresis can be likened to layers of fishing nets placed in the sea to capture a variety of sea creatures. If a suction pump were placed near the nets to suck up the sea creatures, they would move through the nets at different speeds and get caught at different places, depending on the characteristics of each type of creature. For instance, once the pump was turned on, a whale would not be sucked toward it at the same velocity that a sea horse would; and we would expect the sea horse to pass through the openings in the nets far more readily than a whale. In this analogy, the nets represent the gel, the sea creatures (whales, sea horses, jellyfish, anemones, oysters, and so forth) represent the different types of molecules, and the pump represents the electrical field. Just as the sea creatures would move at different speeds and into different places in the nets based on their size, shape, and weight, so do the molecules move through the gel in electrophoresis.

Two types of electrophoresis exist. The first and less expensive is gel electrophoresis. In this type, samples are placed at one end of a slab of polymer gel that has an electrical current passing through it. The electric field pulls the molecules through the gel. Smaller molecules move more easily and quickly. In this technique, molecules separate out largely based on size and are visible as colored bands in the gel. The second and more expensive technique is agarose gel electrophoresis. In this technique, the DNA molecules separate out and migrate through the gel at various rates based on their molecular weights. This allows scientists to examine the DNA segments of an organism.

Spectroscopy

Spectroscopy allows scientists to identify a compound by examining the way it absorbs and scatters electromagnetic radiation. *Scattering* is the redirection of light when it interacts with matter. When the different molecules in a plant compound absorb light, they scatter or emit electromagnetic energy. The emission intensity of an extract can be used to identify and quantify it using a tool called a spectroscope.

In the following lab exercises you will have the opportunity to test your plant for medicinal properties using a variety of different tests, some of which have been described here. You will conduct experiments and draw conclusions from them about your study specimen, just as scientists do across the globe in their search for new medicines.

Unit 5 Questions for Thought

On a separate piece of paper, answer the following questions as thoroughly as possible.

1. How does the use of plants for medicine in the past 50 years compare to how they were used medicinally prior to that?

2. What has helped to increase the life expectancy of Americans in the past 100 years?

3. What are the special characteristics of bacteria?

4. Which bacterial disease are you most familiar with? Describe how it affects the body.

5. If you were testing a plant for its possible antibacterial action, what could you do to see if it inhibits a certain type of bacteria?

6. Why is it so difficult to experiment with viruses and why are there so few effective viral medications?

7. Why is there always a need for new medicines, and why are new strains of old illnesses always developing?

8. Describe the deadliest of the protozoan diseases. Which plant was discovered to help cure it?

9. Why are fungi not considered plants? How do they differ from plants?

10. Explain how fungi can be both helpful and harmful to humans.

11. What is the difference between an infectious and a noninfectious disease?

12. Imagine that you have found a new plant that you have been told may contain powerful medicinal properties. Explain how each of the following procedures could help you understand more about this plant's properties:

   a. petri dish cultures
   b. chromatography
   c. electrophoresis
   d. spectroscopy

Unit 5 Activities

Laboratory Activities: Assessing Potential Medicinal Properties of Your Plant

A. Does Your Plant Have Antibiotic Properties? Part I

Name: ______________________________ Date: ________________________

In this experiment we will find out if your plant extract has the ability to kill *Lactobacillus acidophilus* and/or *L. bifidus*, the bacteria that turns milk into yogurt. These helpful bacteria enhance digestion and are good for you. If your plant is able to kill or inhibit the growth of these bacteria, there is a possibility that it may also be effective in controlling harmful pathogenic bacteria. If so, your plant might be a candidate for further research as a potential antibiotic drug for the future.

Materials needed

Your plant extract (see Unit 3 Activities for instructions on making a plant extract)
One teaspoon of plain all-natural yogurt with active cultures added
One pint of fresh milk
Two small beakers
Two medicine droppers
Masking tape
Thermometer

Directions

1. Put one teaspoon of plain yogurt into a sealable container with one pint of milk. Shake the container so the milk and yogurt mix thoroughly. This is known as *inoculating*, or infecting, the milk with the bacteria.

2. Put 50 ml of the inoculated milk into both of the glass beakers. Use the masking tape to make labels for each jar. Label one of them "Control, no plant extract" and the other one "Experiment, contains [your plant] extract." Also put your initials and the date on each label. If you want to test more plants, add another label and 50 ml to each additional beaker for each plant.

3. Add 5 ml of your plant extract to the experimental beaker and 5 ml of water to the control beaker. Gently swirl the beakers, the same number of swirls each, to mix the contents together.

4. Cover the beakers and put them in a warm place (the warmer the milk, the faster the bacteria will grow). Measure the air temperature where the beakers are located.

5. Every day for the next 5 days, check the beakers and examine how thick the mixture is. The thicker the mixture, the more bacteria has grown in it. Fill in the information below to determine the results of the experiment.

6. When it is over, wash out all of the materials with a 10% bleach solution and/or dispose of the beakers in a tied plastic bag.

Analysis

1. Did you expect your plant to kill or inhibit the growth of the bacteria? What reasons did you have for your hypothesis? ____

____

____

2. Fill in the chart below. If the mixture is thin and watery, write a "1"; if the mixture is a little lumpy or thicker, write "2"; if the mixture turns thick, write "3"; if the mixture turns solid, write "4."

Plant(s)	Thickness of the yogurt after: 1 day	2 days	3 days	4 days	5 days
____________	____	____	____	____	____
____________	____	____	____	____	____
____________	____	____	____	____	____
____________	____	____	____	____	____
____________	____	____	____	____	____

3. Plot your data on the line graph to show how quickly each batch of the mixture thickened.

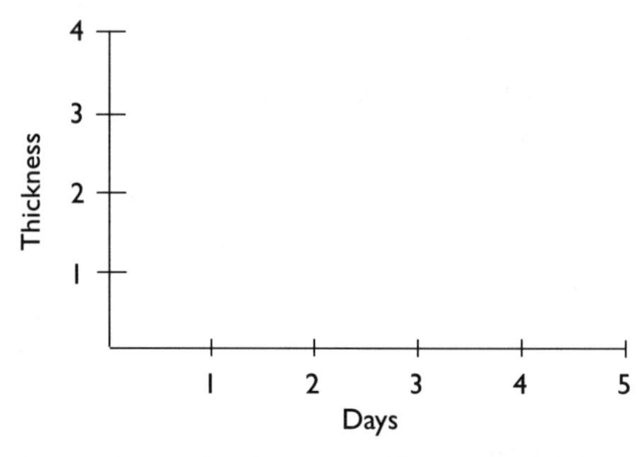

4. How quickly did the mixture in the control beaker solidify? ____________

5. How quickly did the mixture with your plant extract solidify? ____________

6. Did your plant slow down, speed up, completely prevent, or not affect the growth of *L. acidophilus*? ____________________________________

7. What was the air temperature when you set out the milk? ______ How do you think this experiment would change if you were to raise the temperature? What about if you lowered the temperature? ____________________________________

____

____

8. Based on this experiment, do you think your plant might be a good antibiotic?

____

9. Based on this experiment, do you think your plant might be a good food preservative?

____

B. Does Your Plant Have Antibiotic Properties? Part II

Name: ______________________________ Date: ________________________

This experiment demonstrates how bacteria is grown in culture and how substances are tested for their antibiotic properties. This experiment replicates the method by which Alexander Fleming discovered penicillin. If your plant can kill or inhibit bacteria in a petri dish, there is a possibility that it may also be effective in controlling pathogenic bacteria in the body. The plant might be a good candidate for use as a food preservative as well. This would indicate that further research into this plant's antibiotic properties would be beneficial.

Materials needed

Your plant extract (see Unit 3 Activities for instructions on making a plant extract)
Two petri dishes (more, if you want to test more than one extract)
Wax pencil
Sterile cotton swabs
Paper disks or filter paper (preferably sterile) cut into circles with a 2.5 mm radius
Two sterile forceps (or more if you want to test more than one extract)
Sterile nutrient agar

Directions

1. Prepare both petri dishes with nutrient agar. With the wax pencil, write the date and your initials on each one. On one dish write "Control, no plant extract" and the other one "Experiment, contains [your plant] extract." If you are testing more than one plant, write the name of each plant on each dish.

2. Soak one of the paper disks in plain water and the other one in your plant extract (and others in each plant you want to test).

3. Use the sterile swab to rub the inside of your cheek; this will pick up some of the many bacteria that live inside the human mouth. Rub the swab all over the petri dish in a back-and-forth motion in order to completely cover the dish and inoculate it with the bacteria from your mouth. Throw away the cotton swab. Repeat this procedure for the next petri dish.

4. Use a forceps to place the paper disk soaking in plain water onto the control petri dish. Use a different forceps to pick up and place the disk soaking in your plant extract onto the petri dish labeled with the plant extract. Cover the dishes and place them in a warm spot to incubate.

5. Each day check the dishes to observe growth of the bacteria. You can see the bacterial colonies growing as white or yellow patches on the dish. The agar provides the bacteria with food so it can grow. If the plant extract is inhibiting or killing the bacteria and stopping it from growing, there will be a space around the disk where no bacterial colonies are forming. Measure the width of this band with a metric ruler. The band is the "zone of inhibition." It is also possible that your plant stimulates bacterial growth, in which case the bacteria will be growing more densely around and on top of the paper disk.

6. When the experiment is over, discard the dishes in a sealed plastic bag, or autoclave them or clean them with a 10% bleach solution.

Analysis

1. Did you expect your plant to kill or inhibit the growth of bacteria on the petri dish? What reasons did you have for your hypothesis? ___________________________________

____

____

2. Fill in the chart with the results of the experiment, recording the zones of inhibition around each paper disk each day. The band around the disk should be measured in millimeters.

Plant(s)	Zone of inhibition around paper disks after: 1 day	2 days	3 days	4 days
____________________	____	____	____	____
____________________	____	____	____	____
____________________	____	____	____	____
____________________	____	____	____	____
Control	____	____	____	____

3. Did your plant extract inhibit or promote the growth of bacteria around the paper disk?

____

4. Based on this experiment, do you think your plant is a good candidate for more testing as an antibiotic medicine? ___________________________________

5. Based on this experiment, do you think your plant is a good candidate for more testing as a food preservative? ___________________________________

C. Does Your Plant Have Antifungal Properties? Part I: Can It Inhibit Yeast?

Name: ______________________________ Date: ________________________

The purpose of this experiment is to see if your plant extract can inhibit the growth of different types of fungi on bread dough. If it does, then it might be a potential antifungal medication for fungal infections in humans, plants, and/or animals.

In the beginning of this experiment, you will determine whether your plant extract can stop or reduce the rising of bread caused by yeast, which is a single-celled fungus. The type of yeast that makes the bread rise is not harmful to humans. However, if your plant inhibits bread yeast, it may also have applications for controlling pathogenic yeast infections.

Dry yeast is dormant, but when it is added to the moist dough, it begins to digest wheat in the dough. The yeast creates carbon dioxide during its process of metabolizing the food, and it is this process that causes the bread to fluff up and expand with air bubbles. If your plant extract interferes with this process, then the bread will not rise and the yeast will not be able to grow and reproduce. The results of this experiment will be visible within one day.

Materials needed

Your plant extract (see Unit 3 Activities for instructions on making a plant extract)
Two tongue depressors
Two test tubes
Container of water
Plain white flour
Packet of yeast
A wax pencil
Metric ruler
Mixing bowl
Test tube rack
Metric scale
Two spoons
Small plastic plates

Directions

1. Put 250 g of flour in the mixing bowl. Add water slowly until you've added enough to make a dough that is soft but not soggy. If necessary, you can adjust the amounts of water and flour to achieve the right consistency.

2. Use the ruler and wax pencil to mark each test tube 2 cm from the bottom. Near the top of the tube write your initials and the date. On one write "C" for control and on the other write "T" for trial of the plant extract.

3. For each test tube, measure out 10 g of the dough onto a plastic plate. Add 1 ml of water to the control dough and 1 ml of your plant extract to the other dough. (If you are testing more than one plant, add 1 ml of each plant extract to the each batch of dough, and label each one. Use a separate spoon and plate for each batch so that the ingredients will not get mixed together.)

4. Mix 25 mg of yeast and knead it into each batch of dough. Add a little more flour to any of the doughs that are very watery, so that they become less gooey and a little firmer.

5. Quickly, before the yeast begins to make the dough rise, use a stick to put a small amount of the control batch of dough into the control test tube. Use enough dough to reach the 2 cm mark when you press it down firmly.

6. Use another stick to put some of the dough containing your plant extract into the test tube labeled "T"; make sure it reaches the 2 cm mark when you press it down firmly. (If you plan to test additional plant extracts, repeat this procedure for each one.)

7. Discard the remaining dough and clean all the dishes.

8. After 30 minutes, and again after 24 hours, measure each batch of dough in the test tubes to determine how much the dough has risen from its original 2 cm mark. Record your results on the chart below.

9. When the experiment is finished, discard the test tube and their contents in a sealed plastic bag, or autoclave them or wash them with a 10 % bleach solution.

Analysis

1. Did you expect your plant to inhibit the growth of fungi? What reasons did you have for your hypothesis? ______________________________

____

____

2. In the chart below, record the amount that the bread dough rose in each of the test tubes, in centimeters.

Plant(s)	Amount the bread dough rose after: 30 minutes	24 hours
____________	______ cm	______ cm
____________	______ cm	______ cm
____________	______ cm	______ cm
____________	______ cm	______ cm
Control	______ cm	______ cm

3. In the space below, create a bar graph to demonstrate how much of each batch of bread dough rose. Make sure to label the graph, the heights, and each bar.

4. Did your plant extract prevent, slow down, speed up, or make no difference in the rising of the bread dough? ______________________________

D. Does Your Plant Have Antifungal Properties? Part II: Can It Inhibit Bread Mold?

Name: ______________________________ Date: ________________________

In this experiment you will determine whether your plant extract inhibits or encourages the growth of mold on bread. Mold is another kind of fungus that helps organic materials to rot and decompose. As in Part I of this section on antifungal properties, if the mold does not grow on the bread, or if it grows more slowly with your plant extract, your plant may be inhibiting the growth of the fungi.

Materials needed

Your plant extract (see Unit 3 Activities for instructions on making a plant extract)
Water
Two medicine droppers
A plate
Two slices of plain white bread
Two sealable plastic bags
Marker pen

Directions

1. Place the bread slices side by side on a plate and leave them in the open for at least 15 minutes so they will be exposed to mold spores present in the air.

2. With the marker, write "Control: water added" on one bag; on the other bag write "Test: [your plant extract] added." Write your name and the date on both bags. Use one medicine dropper to place 20 drops of plain water in the center of one of the bread slices. Place this slice in the bag labeled "Control" and seal the bag.

3. Use the other medicine dropper to place 20 drops of your plant extract in the center of the remaining bread slice. Place this slice in the bag labeled "Test" and seal the bag.

4. If you want to test more than one plant extract, prepare a bag and bread slice as in steps 1 and 3. Be sure to use a clean medicine dropper for each extract you test.

5. Place the sealed bags in a warm spot.

6. Observe the bread in the plastic bags approximately every other day for two weeks, watching for the formation of bread mold.

7. When the experiment is complete, do not open the sealed bags; dispose of the bread and mold sealed in the plastic bags. (**Note:** Mold spores can trigger or exacerbate allergies.)

Analysis

1. Did you expect your plant to inhibit the growth of the mold? What reasons do you have for your hypothesis? ____

____

____

2. In the chart below, record the amount and color of the mold growing on the centers of the bread slices over the two-week period. If there was no mold growing, write "1"; if there was just a little, write "2"; if there was a lot of mold, write "3"; if the center of the bread was totally covered in mold, write a "4." Use the same numbers to record the amount of mold on the edges of the bread.

Name of plant extract: ______________________________

Dates	_____	_____	_____	_____	_____	_____	_____
Control:							
Center	_____	_____	_____	_____	_____	_____	_____
Edges	_____	_____	_____	_____	_____	_____	_____
Mold color	_____	_____	_____	_____	_____	_____	_____
Test:							
Center	_____	_____	_____	_____	_____	_____	_____
Edges	_____	_____	_____	_____	_____	_____	_____
Mold color	_____	_____	_____	_____	_____	_____	_____

3. Plot your data on the following line graph, using separate lines for the center and edges of the bread.

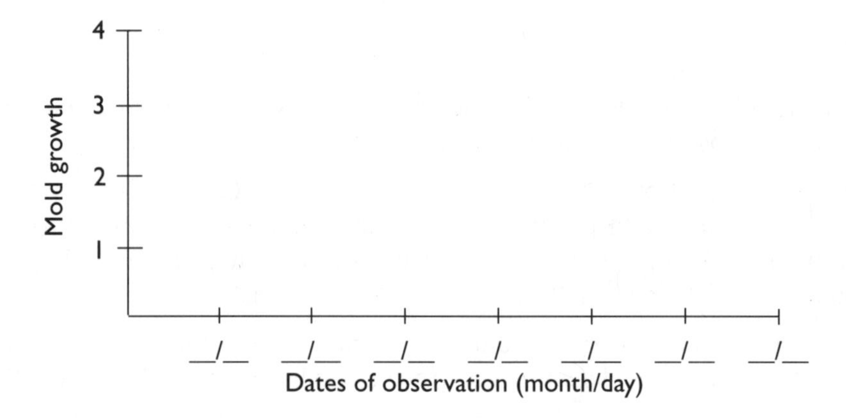

4. What effect did your plant extract have on the growth of bread mold?

____

5. Based on this experiment, do you think your plant extract has the ability to inhibit mold? ____

E. Does Your Plant Have Antiprotozoal Properties?

Name: ______________________________ Date: ________________________

In this experiment you will grow harmless protozoans and see the effect of your plant extract on them. Some protozoans live in pond water and can be collected from ponds or streams, or they can be purchased through a biological-supply company. Sometimes they come in a dry mix and water can be added to bring them out of dormancy. Amoeba, paramecium, arcella, euglena, chilomonas, stentors, and volvox are good choices of protists to conduct this experiment on. They can live for several weeks or months in a classroom in clean water to which small amounts of algae, straw, and soil are added as food sources. If your plant extract can inhibit or kill these harmless protozoans, it might have potential for use in combating disease-causing protists such as the plasmodium that causes malaria.

Materials needed

A dissecting, stereo, or field microscope; or a powerful magnifying glass
Specimens of living, nonharmful protozoans
Two petri dishes (plus an additional one for each additional plant you wish to test)
Clear cellophane tape
Fine-pointed marker pen
Two medicine droppers
Metric ruler
Your plant extract (see Unit 3 Activities for instructions on making a plant extract)
Wax pencil

Directions

1. Use the wax pencil to write your initials and the date on the lid of each petri dish. On one lid write "Control, no plant extract" and on the other write "Experiment, [name of your plant] extract."

2. Draw a 0.5 cm square on a piece of clear tape with the marker, and affix the tape to the underside of the petri dish, in the center.

3. Using a medicine dropper, add the protozoan/water mixture to each petri dish, to a level of 0.25 cm.

4. Use the microscope or magnifying glass to locate living protozoans in your petri dish. Count only those protozoans found during a two-minute period within the marked square you placed in the dish. Record the number of microbes seen in each dish on the chart on the following page. (**Note:** If the protozoans in your dishes are not living or healthy, you will need a fresh stock of healthier organisms to complete this experiment.) Identify the types of protozoans you found, using the drawings of protozoan types on page 68. Record how active they are.

5. Using the other medicine dropper, put two dropperfuls of your plant extract into the experimental petri dish. Add two dropperfuls of plain water to the control dish. Cover both.

6. After 30 minutes, examine the protists under the microscope again. Observe the marked square for two minutes and record for each dish the number of protists and the activity level.

7. After 24 hours examine the protists under the microscope once again, using the same procedure as in the previous step.

8. After 48 hours, repeat step 6.

	Control	With plant extract
Number of live protozoans at start	_______	_______
Type of protozoans	_______________	_______________
Health of protozoans	_______________	_______________
Number of live protozoans after30 min	_______	_______
Type of protozoans	_______________	_______________
Health of protozoans	_______________	_______________
Number of live protozoans after 24 hr	_______	_______
Type of protozoans	_______________	_______________
Health of protozoans	_______________	_______________
Number of live protozoans after 48 hr	_______	_______
Type of protozoans	_______________	_______________
Health of protozoans	_______________	_______________

Analysis

1. How did your plant extract affect the protist colony? ______________________________

____

____

2. Based on this experiment, do you think your plant might be a good candidate for further testing as an antiprotozoal medicine? Why, or why not?

____

____

3. Make a line graph from the data you recorded to illustrate the number of protists counted from the beginning to the end of the experiment for both the control and the experimental test with your plant extract. Make one line for the control and one for the test using another color, and make a legend to distinguish which line is which.

F. Does Your Plant Have Cytotoxic Properties?

Brine shrimp (*Artemia salina*) are small, easy-to-culture animals that are sometimes used to test the cytotoxicity potential of plants. If your plant extract kills the brine shrimp, it is possible that it might also be able to be used to kill cancer cells. In this experiment, you will grow some brine shrimp and then apply your plant extract to different samples to see how it affects the tiny shrimp.

Brine shrimp kits can typically be found a pet store or a biological-supply company. (**Note:** Most states allow students to experiment with invertebrate animals such as brine shrimp in such a way that may cause mortality to the animals. But you should check with your teacher and local school board to ensure that this procedure is acceptable in your state and school district.)

Materials needed

Brine shrimp kit with eggs or adult brine shrimp, salt solution to emulate sea water, food for the shrimp (they like baking yeast), and a container
Dissecting or stereo microscope
Your plant extract (see Unit 3 Activities for instructions on making a plant extract)
Three petri dishes
Wax pencil
Small spoon
Medicine dropper
Metric ruler

Directions

1. Following the directions on the kit, grow your brine shrimp colony; when you have a healthy colony, you are ready to begin the experiment.

2. Prepare the petri dishes by measuring and marking the halfway point of each dish's depth. Label each dish with your initials and the date. On the control dish write "Control." On the second dish, write "10 drops of [plant name] extract" and write "20 drops of [plant name] extract" on the third dish.

3. Using a medicine dropper, fill each petri dish to the halfway mark with the brine shrimp solution. Select about 10 healthy brine shrimp to place into each dish, using the spoon and taking care not to injure them while transporting them. It might also help to observe them under the dissecting microscope while you are catching them to make sure you have about 10 healthy shrimp in each dish.

4. In the control dish, use the medicine dropper to add 20 more drops of the brine shrimp solution.

5. In the dish marked "10 drops," add 10 drops of your plant extract and 10 drops of brine shrimp solution.

6. In the dish marked "20 drops," add 20 drops of your plant extract.

7. Mix the contents of each dish by gently swirling each one, the same number of swirls for each dish.

8. After 24 hours, observe the brine shrimp in each petri dish under the dissecting microscope and record how many are thriving, how many seem sick or sluggish, and how many have died. Record your data in the chart in the Analysis section, on the next page.

Analysis

1. What was your plant extract? ________________________________

2. Fill in this chart with the results of your experiment. Convert your numbers into percentages. For example, if 3 out of 10 brine shrimp were dead, then 30% died; if 2 out of 8 died, then 25% died.

	Dead brine shrimp	Sluggish brine shrimp	Healthy brine shrimp
Control	______%	______%	______%
10 drops	______%	______%	______%
20 drops	______%	______%	______%

3. Based on this experiment, do you think your plant extract might have strong cytotoxicity potential? Give reasons for your answer. ________________________________

____

____

____

4. If you or your classmates tested more than one plant extract, which one had the greatest cytotoxicity action? ________________________________

Laboratory Activity: Paper Chromatography Test

Name: ______________________________ Date: ________________________

This experiment will allow you to separate some of the pigments hidden in your plant. The following information may help you identify some of the compounds you see:

Chlorophyll A is bright green
Chlorophyll B is a dull or khaki green
Carotenoids are lemon yellow to orange
Anthocyanin is pale pink, cherry red, purple, or blue
Xanthophyll is yellow

The first part of the experiment involves trying out paper chromatography with black markers so as to get practice before testing on your plant.

Materials needed

Two different brands of black markers, labeled "A" and "B"
Your plant tincture (and other tinctures you would like to try; see Unit 3 Activities for instructions on making a tincture of your plant)
Pieces of coffee filter paper cut into strips 1 cm wide and 12 cm long
Test tubes
Solvent made of 20 ml water, 20 ml rubbing alcohol, and 5 ml vinegar
Metric ruler
One medicine dropper for each tincture you will test
Pencil
Test tube holder

Directions

1. Make a tiny pencil mark 2 cm from the bottom of the filter paper. This is where you will place your pigments.

2. With the first marker make a dot, about the size of a pencil eraser, on the tiny pencil mark you made on a filter strip. Repeat this several times on the very same spot so that more ink is dispersed onto the paper. Do the same with the second marker, on another filter strip. At the top of each strip write your initials, and label one strip "A" and the other "B" to correspond with the labeled markers.

3. For each of the filter papers, prepare a test tube by pouring in the solvent until it reaches 1 cm high.

4. Place each of the marked filter papers into a separate test tube. Press the papers down to the bottom of each tube so the solvent can be absorbed by the paper but does not touch the pigment. Keep the test tubes upright in a test tube holder.

5. For each plant tincture you wish to test, use a separate medicine dropper and strip of filter paper. After repeating step 1 with the new paper strip(s), place a drop of the plant tincture on the tiny pencil mark you made. Label the top of each paper with the name of the plant tincture being used, and write your initials.

6. Let the drop of tincture dry, then place more tincture directly over it. Repeat this step four times.

7. Make a control filter with no pigment on it. Label it "Control" and write your initials at the top.

8. For each of the tincture strips and the control strip, repeat steps 3 and 4.

9. Observe the filter paper as the pigments begin to separate in each test tube. It usually takes 20 to 60 minutes for the pigments to fully separate.

10. Once the pigments have fully separated, take them out of the test tubes and place them on a sheet of paper or a tray to dry. Record your observations below.

Analysis

1. Tape the filter papers to your paper as soon as they dry out. Measure the distance from the original ink or pigment spot to the approximate middle of each streak of color, to see how far each compound traveled.

Marker A Color	Distance	Marker B Color	Distance
____________	____ cm	____________	____ cm
____________	____ cm	____________	____ cm
____________	____ cm	____________	____ cm
____________	____ cm	____________	____ cm
____________	____ cm	____________	____ cm

Control Color	Distance	My plant tincture: ____________ Color	Distance
____________	____ cm	____________	____ cm
____________	____ cm	____________	____ cm
____________	____ cm	____________	____ cm
____________	____ cm	____________	____ cm
____________	____ cm	____________	____ cm

Other plant tincture: ____________ Color	Distance	Other plant tincture: ____________ Color	Distance
____________	____ cm	____________	____ cm
____________	____ cm	____________	____ cm
____________	____ cm	____________	____ cm
____________	____ cm	____________	____ cm
____________	____ cm	____________	____ cm

2. Can you identify the pigments you found in your plant based on the color? If so, what are they? ____

____

3. What do you think determined how fast and how far each pigment travels?

____

____

Laboratory Activity: Gel Filtration Chromatography Test

Name: ______________________________ Date: ________________________

Gel filtration is a more sophisticated chromatography method that is used by many biotechnology companies to separate substances such as proteins in the development of new drugs. The gel beads are the stationary medium through which the solvent moves. The smaller molecules penetrate the gel while the larger molecules pass through faster and move farther.

Materials needed

Your plant tincture (see Unit 3 Activities for instructions on making a tincture of your plant)
Small chromographic columns (one for each plant tested) with gel beads (try to purchase gel columns that are about the size of a test tube and have caps at the top and bottom)
Ethyl alcohol
15 ml graduated cylinder
Two medicine droppers
Metric ruler
Test tube brush
Colored pencils
Wax pencil

Directions

1. Put some ethyl alcohol through each column and let it run through the filter, to prime the gel beads.

2. When the alcohol empties out, cap the bottom of each column. Measure and mark the column three-quarters of the way from the bottom.

3. Add the wet gel into each column until the column is three-quarters full.

4. Add more alcohol and let the gel settle.

5. Remove the cap so excess alcohol can drain out.

6. Use a medicine dropper to add 10 drops of your plant tincture to one of the columns. Do the same for each additional plant tincture you wish to test (with a sterile dropper for each), and label each column with the name of the plant tincture added to it.

7. Observe the column as the tincture travels through it and the different pigments separate. Record your data on the chart below.

Analysis

1. Measure the distance from the top of the gel column where the plant was first applied to the approximate middle of each streak of color in the gel column. Record your data below.

My plant tincture: _______________		Other plant tincture: _______________	
Color	Distance	Color	Distance
____________________	____ cm	____________________	____ cm
____________________	____ cm	____________________	____ cm
____________________	____ cm	____________________	____ cm
____________________	____ cm	____________________	____ cm
____________________	____ cm	____________________	____ cm

2. Can you identify the pigments you found in your plant, based on their color? If so, what are they? ____________________

____

____

3. What do you think determines how fast and how far each pigment travels?

____

____

4. Were the pigments that separated out from your plant the same in the gel chromatography experiment as the ones in the paper chromatography experiment? Were the pigments layered in the same configuration? ____________________

____

____

Summary of the Medicinal Experiments

Name: ______________________ Date: ______________________

Plant name: ______________________ Part(s) used for testing: ______________

Use this page to write a summary of what you have learned about the medicinal properties of your plant. If you would like to share these results with other interested students, teachers, and scientists on the World Wide Web, contact the author at gdpaye@hotmail.com for instructions on how to do so.

1. Did my plant kill or inhibit *L. acidophilus* bacteria in yogurt?

	Thickness of the yogurt after:				
Plant(s)	1 day	2 days	3 days	4 days	5 days
______________	____	____	____	____	____
______________	____	____	____	____	____
______________	____	____	____	____	____
______________	____	____	____	____	____

What was the air temperature when you set out the milk? ______

2. Did my plant inhibit bacteria growing in a petri dish?

	Zone of inhibition around paper disks after:			
Plant(s)	1 day	2 days	3 days	4 days
______________	____	____	____	____
______________	____	____	____	____
______________	____	____	____	____
______________	____	____	____	____
Control	____	____	____	____

3. Did my plant inhibit growth of yeast and mold on bread dough?

	Amount the bread dough rose after:	
Plant(s)	30 minutes	24 hours
______________	______ cm	______ cm
______________	______ cm	______ cm
______________	______ cm	______ cm
______________	______ cm	______ cm
Control	______ cm	______ cm

4. Did my plant inhibit or kill the protists?

	Control	With plant extract
Number of live protozoans at start	______	______
Type of protozoans	________________	________________
Health of protozoans	________________	________________
Number of live protozoans after30 min	______	______
Type of protozoans	________________	________________
Health of protozoans	________________	________________
Number of live protozoans after 24 hr	______	______
Type of protozoans	________________	________________
Health of protozoans	________________	________________
Number of live protozoans after 48 hr	______	______
Type of protozoans	________________	________________
Health of protozoans	________________	________________

5. Which pigments do my plant contain?

From the paper chromatography test:

My plant tincture: ____________________________________

Color	Distance
________________________	_____ cm
________________________	_____ cm
________________________	_____ cm
________________________	_____ cm
________________________	_____ cm

From the gel filtration chromatography test:

My plant tincture: ____________________________________

Color	Distance
________________________	_____ cm
________________________	_____ cm
________________________	_____ cm
________________________	_____ cm
________________________	_____ cm

Other plant tincture: ____________________________________

Color	Distance
________________________	_____ cm
________________________	_____ cm
________________________	_____ cm
________________________	_____ cm
________________________	_____ cm

Based on the colors, can you identify the pigments you found in your plant? If so, what are they? ______________________________

____

____

6. Did my plant kill or inhibit the following:
   - Acidophilus bacteria in yogurt? ______________________________
   - Bacteria growing in a petri dish? ______________________________
   - Yeast in bread dough? ______________________________
   - Protozoans in pond water? ______________________________
   - Brine shrimp? ______________________________

7. Which plant pigments did my plant have in the paper chromatography lab?

____

Which pigments did it have in the gel filtration chromatography lab?

____

8. Based on these experiments, what have you learned about the medicinal properties of your plant? ______________________________

____

____

____

____

____

____

____

Unit 6
Testing Your Plant for Other Useful Properties

In Units 4 and 5 we learned about the importance of plants as food and medicine. In this unit we will explore how people use plants for a multitude of other uses including the production of paper, insecticide, fertilizers and plant growth stimulants, animal feed, dyes, fresh cut flowers or foliage, and fragrances. The experiments presented in this chapter will help you discover whether your own plant might furnish any of these products. You may even discover some additional uses along the way.

Products from Plant Fibers

Cotton fabric, natural broom bristles; woven door mats, hats, and baskets; carpet backing; rope and twine; stuffing materials; and, of course, paper — these are some of the products from plant fibers that we use in our everyday lives. They are derived from several different plant parts including fruit and seeds (cotton, kapok, and coir from coconut); soft stem fibers (jute, flax, and industrial hemp); stiff leaf fibers (palm, sisal); and the fibers in wood. Although plant fibers are derived from a large variety of unrelated plants, they can be divided functionally into two groups: (1) relatively long, flexible fibers that can be woven, plaited, or spun into thread; and (2) short, brittle, and/or slippery fibers such as those in wood pulp that we use to make paper. One of the activities in this unit is to try to make paper from your own study plant.

The Egyptians are usually credited with making the first paper because 5,500 years ago they wrote on papyrus leaves (*Cyperus papyrus*) that had been pressed together. The name

A Plant of Many Virtues

Every year we use three billion cubic meters of wood worldwide to construct buildings, boats, furniture, and fences. Current logging practices are exceeding the ability of natural forests to regenerate. The increased use of plants such as bamboo can help forest conservation. More than 800 species of bamboo (Bambusaceae family) have been identified; taken together, they yield more than 1,000 different products from their stems and leaves.

Bamboo grows incredibly fast. Some species grow taller than many trees, and bear stems of greater strength than many types of timber. The stronger stems are used to construct houses, bridges, and scaffolding; smaller stems are used to craft furniture. Bamboo fibers can be spun to make rope and twine, the leaves used to make roof thatch, and young tender shoots eaten as a vegetable. In Asia, the bamboo symbolizes virtue, humility, and resistance to hardship.

"paper" is derived from the word "papyrus." Other cultures, such as the Mayans of Central America and the Polynesians, have also made paperlike substances by pounding sheets of bark from plants in the mulberry family (Moraceae).

Paper as we know it today is formed from plant fibers that have first been separated from one another in a slurry and then matted together again in a thin sheet. The Chinese began making paper this way by 200 B.C., using fibers from the paper mulberry tree mixed with fibers from hemp (*Cannabis sativa*) and flax (*Linum usitatissimum*). It was not until 1,000 years later that paper-making spread to Europe. The first paper in Europe, however, was made from rags rather than wood. The idea of using wood pulp as a paper source came from a French inventor in the early eighteenth century who was inspired by watching wasps building papery nests from wood.

Paper making is an energy-intensive process in which wood chips are dissolved into a pulp by chemical and/or mechanical means. The pulp is fed onto a screen of wire mesh, drained, pressed by rollers, and ironed. In the final product the fibers have interlocked, forming sheets of paper.

Today the majority of the world's paper supply comes from coniferous forests and forest plantations in the United States, Canada, and Scandinavia. Each year a forest region covering about 450,000 square kilometers (174,000 square miles) — roughly the size of Sweden — must be felled to supply the world's paper needs. The United States has the highest rate of paper consumption in the world. In 1990 we used almost 320 kilograms (700 pounds) per person, with most of it ending up in landfills. An increase in paper recycling would help reduce pressure on our dwindling forest resources. However, the problem with recycling is that each time the paper is recycled, the fibers become shorter. New sources of plant fibers must be added to maintain paper quality.

Paper that Doesn't Grow on Trees

The woody herb kenaf (*Hibiscus cannabinus*) is now being grown commercially in southern Texas and Louisiana as an alternative source of paper pulp. One acre of kenaf yields five times the amount of pulp that an acre of pine trees can supply. Plants such as these will help lessen our dependence on old-growth forests or forest plantations as a source of paper.

Cotton (*Gossypium* species), jute (*Corchorus* species), industrial hemp (*Cannabis* species), flax, and kenaf (*Hibiscus cannabinus*) are all examples of non-woody plants currently used to make paper. The *manila envelope* is named for Manila hemp (*Musa textilis*), a relative of the banana. Fibers taken from the leaf bases of this plant have been used to manufacture manila envelopes, tea bags, dollar bills, and casings for salami. Many countries use agricultural wastes to make paper: for instance, China uses large amounts of rice straw; India, Brazil, Mexico, and China also take advantage of *bagasse*, a waste product left over from the processing of sugar cane. Perhaps you will find that your study plant contains fibers that could be successfully used as a component in new or recycled paper. (At left, Amity is in the process of making her own paper, as you will have the opportunity to do in the activities for this unit.)

Applications for Gardening and Agriculture

In this section we discuss the use of plants as an aid to agriculture, and point out how they tie into the experiments presented at the end of this unit. We discuss the use of plants as insecticides, as fertilizers or growth stimulants, and as animal feed. A brief discussion on pH is also presented because of its importance to plant growth.

Plants that Control Pests

We are currently suffering a number of environmental problems that are related in part to contamination of our soil, food, and water systems by agricultural chemicals. Although DDT (dichloro-diphenyl-trichloro-ethane) has been banned from use in the United States for many years now, it persists in the environment, moving up through the food chain, accumulating in the fatty tissues of living organisms all over the world. For example, some researchers have found DDT and its derivatives in the tissues of animals in the polar regions. DDT is also present in significant quantities in human breast milk throughout the world. This family of pesticides is now known to affect the endocrine and reproductive systems, and can cause cancer.

Most commercial pesticides affect a broad range of insects, killing off beneficial insects as well as those that cause problems. When the populations of natural predators are killed off along with the pests, the stage is set for recurring problems. Pests can quickly repopulate in an area free of natural predators that originally kept their population in balance. Furthermore, the frequent use of synthetic pesticides promotes the evolution of insects that are resistant to the chemicals being used to control them.

Just as they have potential to supply us with yet-undiscovered medicines, plants can also be sources of new pesticides. Having coevolved with insects, plants have developed an entire arsenal of chemicals and other defenses to protect themselves from insect predators. By exploring the potential of plants to produce natural insecticides, we are tapping into this rich evolutionary treasure chest. One such jewel is the plant *Chrysanthemum cinerariifolium* (Asteraceae), the source of pyrethrum. Pyrethrum is an insecticide that stuns insects, though it does not instantly kill them. The active compounds in pyrethrins have been incorporated into a significant number of commercial pesticides worldwide. Large-scale production of pyrethrum for the international market is found in regions of Kenya, Tanzania, and South America.

Derris eliptica, a plant used as a fish poison used in Amazonia, was found to be a source of rotenone. Rotenone is an effective insecticide that is completely harmless to people because it leaves no residue. Likewise, the neem tree (*Azadirachta indica*) from India also contains extremely potent compounds with insecticidal properties that have only recently been discovered. Farmers in India have traditionally used leaves of the neem tree in their grain storage bins to kill weevils. Neem leaves are toxic to insects but not to people. Both neem and rotenone are good examples of effective pesticides discovered through ethnobotanical leads.

It is important to note that although a pesticide may be "natural" or plant-derived, that does not necessarily make it safe. Pesticides derived from the tobacco plant, for instance, are extremely poisonous.

Many plants exist that repel insect pests although they have not been developed as commercial pesticides or repellents. Organic gardeners often spray their vegetable plots with various teas or decoctions made of plants such as cayenne pepper (*Capsicum annuum*), lemongrass (*Cymbopogon citratus*), pennyroyal (*Mentha pulegium*), rosemary (*Rosmarinus officinalis*), sage (*Salvia officinalis*), thyme (*Thymus vulgaris*), wormwood (*Artemisia absinthium*), and garlic (*Allium sativum*). Hot or strong-tasting plant sprays may also ward off vertebrate pests such as rats, rabbits,

cayenne pepper (*Capsicum annuum*)

eucalyptus (*Eucalyptus globulus*)

and raccoons. People (and their pets!) can use certain plant-derived products to protect themselves from mosquitoes, ticks, chiggers, fleas, and lice; those most commonly used are the essential oils of pennyroyal, rue (*Ruta graveolens*), citronella (*Cymbopogon nardus*), rosemary, and eucalyptus (*Eucalyptus* species). To protect our clothing, moth balls were traditionally derived from camphor trees (*Cinnamomum camphora*); cedar wood (*Cedrus* species) also repels moths.

COMPANION PLANTING. Another way to take advantage of the protective benefits one plant might lend to another is through *companion planting.* This practice of combining two types of plants in the garden may offer protection from pests through several mechanisms. First, one plant may produce a natural repellent that protects its neighbor plant from pests. Second, a companion plant species may provide shelter for beneficial insects. Third, combining different kinds plants together in a garden or agricultural plot creates a more diverse ecosystem, making it less susceptible to attack.

Some examples of companion plants that repel insects include (1) mint (*Mentha* species), to protect cabbage (*Brassica oleracea*) from damage by flea beetles and caterpillars; (2) nasturtiums (*Rorippa* species), to repel beetle pests from pumpkins and squash (species of *Cucurbita*); and (3) marigolds (*Tagetes* species), to protect any vegetable plant from root damage by nematodes. Selecting appropriate plants for the garden, including companion plants, will vary according to different climates and soil types. In this respect, searching the Internet for gardening information specific to your region is the most effective way to find out more details on companion planting and natural methods of insect control.

Plants as Fertilizers

Organic matter from decayed plant tissue adds nutrients to the soil just as commercial fertilizers do. Organic matter also improves soil's water-holding capacity, builds soil structure, and provides a source of food for beneficial soil organisms such as earthworms. The amount of nutrients added depends on the source and quantity of organic matter added to the soil.

Many ways exist of adding organic matter to the soil including the addition of prepared compost, the incorporation of a green manure crop, and the use of mulch on the soil surface.

COMPOSTING: Composting is a way of taking advantage of organic waste products such as manure, household vegetable debris, lawn clippings, and leaves. If these materials are set aside in a selected site apart from the garden plot or agricultural field, and are given time to break down into their natural components, they become the valuable source of fertilizer known as *compost.* Composting also helps reduce the impact we have on landfills. On average, Americans throw away nearly 700 kg (1,500 lbs) of trash per person each year. People who compost throw away an average of only 170 kg (375 lbs) of trash per person per year.

GREEN MANURE: A green manure crop is a crop that is grown with the specific purpose of turning or plowing the plant material into the soil while still green so that it decomposes and releases the nutrients directly into the field. Green manure crops form an important part of crop rotation schemes used on organic farms to increase soil fertility without using chemical fertilizers. Among the plants that make especially good fertilizers are many of those in the bean family (Fabaceae) — such as alfalfa (*Medicago sativa*), clover (*Trifolium* species), and peanuts (*Arachis hypogaea*) — because they are "nitrogen-fixers." This means they have nodules on their roots that contain bacteria able to convert elemental nitrogen (N_2) into a form usable by plants. Since the lack of available nitrogen in the soil is often the most limiting factor to crop growth, this "free" addition of nitrogen is noteworthy.

MULCH: A mulch is any material such as straw, grass cuttings, leaves, wood chips, pine needles, and even newspaper spread on the surface of the soil to help reduce the loss of soil moisture, control soil temperature extremes, and suppress weeds. An added benefit is that as the organic matter decomposes, it adds nutrients and organic matter to the soil. Compost can also be used as a mulch by laying it on top of the soil (between the rows of crop plants) rather than turning it into the soil.

The three minerals that plants need in the largest quantities from the soil are nitrogen (N), phosphorus (P), and potassium (K). Nitrogen promotes aboveground growth and imparts a deep green color to the leaves, phosphorus promotes root growth, and potassium is necessary for adequate flowering and fruit development. Commercial fertilizers contain these three minerals in varying percentages. For example, a fertilizer marked as 5-10-5 contains 5% nitrogen, 10% phosphorus, and 5% potassium. In large-scale agriculture, nitrogen is usually the nutrient added in the greatest quantity. One of the problems with the heavy use of chemical fertilizers in commercial agriculture, however, is the runoff or leaching of applied nitrates from the fields into our aquatic systems. Nitrates contaminate our drinking water and damage fish habitat. An increased use of organic sources of fertilizer and a decrease in the use of chemical nitrates would alleviate this problem.

The photograph on the first page of this unit shows Quan with his study on the use of rice water as a ferilizer; in one of the laboratory activities in this unit, you will consider your plant's potential to serve as an organic source of fertilizer. You will conduct an experiment in which you add your plant extract to seedlings growing in pots and observe how the extract affects their development. You will compare that effect with the effect produced by commercial fertilizer on a similar group of seedlings. If your plant extract stimulates growth in the seedlings, it may be due to the nutrients added in the liquid. Hence, your plant extract would be functioning as an organic fertilizer.

In actual field trials, your plant could be used as a source of fertilizer by being added to the garden as compost, as mulch, or even as a green manure crop. In the following chart you will find a list of some common sources of organic matter that are often added to garden soil. (The levels of nitrogen, phosphorus, and potassium are listed in particular because these three minerals are needed in comparatively large amounts for healthy plant growth, and they are commonly found in commercial fertilizers.) Note the high level of nitrogen contained in the alfalfa and soybean meal, both members of the bean family.

Plant material	Nitrogen (%)	Phosphorus (%)	Potassium (%)
Alfalfa hay	2.5	0.5	2.0
Brewers grains	0.9	0.5	0.1
Coffee grounds	2.0	0.4	0.7
Cottonseed	3.0	1.3	1.2
Decomposed leaves	0.6	0.2	0.4
Rotted grass/weeds	2.0	1.1	2.0
Red clover	0.6	0.1	0.5
Seaweed	1.7	0.8	5.0
Soybean meal	6.0	1.2	1.5
Wood ashes	0	1.5	7.0

Keep in mind that your plant extract might also affect seedling growth due to some other stimulating (or inhibiting) compound in the extract. Seedling growth might even be affected by the pH of the liquid added.

PH AND PLANT GROWTH. The pH of a substance refers to its acidity or alkalinity. You can use a pH test strip or a pH meter to determine the pH of your plant extract. Pure water has a pH of 7.0; it is neutral and contains a balance of H^+ and OH^- ions. When pH measures 6.9 or less, it is acid and contains a greater quantity of H^+ ions. Lemon juice and vinegar are examples of common acids. When pH is 7.1 or higher, it is a base and contains more OH^- ions. Baking soda, wood ashes, and limestone are examples of common bases. When a base and acid mix, they neutralize each other. The H^+ and OH^- ions form water molecules, H_2O.

Most plants grow best in soils with a pH ranging between 5 and 7. A pH that is too high or too low affects nutrient absorption and plant growth. At pH extremes, either too acid or too basic, some minerals become toxic. For example, in very acid soils, iron and aluminum become so soluble that they are toxic to plant growth. The pH of your plant extract may have an overall effect on plant growth. Your lab test will help you decide.

Plants as a Source of Animal Feed

Plants are food not only for people but also for animals. A large part of our agricultural land, especially that used to grow corn (*Zea mays*) and soybeans (*Glycine max*) in the United States is dedicated to raising feed for livestock. In this unit you will have the opportunity to test your plant as an animal feed on an animal of your choosing.

A fun and interesting experiment is to see how pets react to your plant as a food source. Even carnivores such as dogs have some of their nutritional needs supplied by plants. Garlic has been noted as beneficial for some pets, cleansing the digestive system and helping to repel parasites. Seaweed adds iodine and other trace elements to a pet's diet; sources of whole grains may provide vitamin E and the B-complex vitamins too.

Before feeding any plant material to an animal, it is essential to learn about that species' nutritional requirements. For example, guinea pigs, like humans, cannot synthesize their own vitamin C and require a dietary source. Guinea pigs will develop scurvy if not provided with foods such as cabbage, oranges, or fortified pellets. You also need to make sure the plant you are studying would not be poisonous to the animals involved in your test. Never treat animals cruelly in any experiment.

You can experiment with feeding a variety of seeds and berries to birds in a feeder. Make observations on different species and their food preferences.

Caution: Do not feed wild animals (other than birds with a bird feeder), since they may carry rabies or other diseases.

corn (*Zea mays*)

You Can Even Eat the Forks!

The soybean (*Glycine max*), the most economically important member of the bean family, is prized for many reasons. Soybeans contain one of the highest levels of useful protein of any plant food. The Chinese name for soybeans translates into "cow-without-bones."

Soybeans have been developed into a multitude of meat and dairy substitutes including tofu, tempeh, texturized vegetable protein, soy dogs, soy burgers, soy cheese, soy milk, and infant formula. Eating soyfoods may reduce the risk of heart disease and cancer due to the presence of a group of phytochemicals known as *isoflavones*.

Other important soy products include soybean oil, soy sauce, dog food, cattle feed, and a host of industrial products including fertilizer, paints, adhesives, paper, engine fuel, and stabilizers for dusty roads.

The versatile soybean just sprouted another amazing use: the production of "edible" forks and spoons. Developed by researchers at the University of Iowa, this new plastic is made from completely biodegradable soy proteins. The soy-based tableware is being tested by the U.S. Navy, now under orders to reduce pollution at sea. No more permanent plastic plates and utensils overboard — these spoons and forks can be ground up and tossed into the sea as a nutritious snack for fish.

Plants as Ornamentals

Since earliest recorded history, people have used plants to adorn our homes and gardens. In modern urban environments, ornamental plants used both indoors and outdoors often remain people's only link with the natural world. Given our passion for using for plants to enhance the beauty of our surroundings, ornamental horticulture and landscape design have grown into multibillion-dollar industries.

Apart from their beauty, plants also enhance the environmental health of indoor and outdoor ecosystems. Certain houseplants such as the spider plant (*Chlorophytum comosum*) absorb indoor air pollutants, such as formaldehyde, that aggravate asthma and allergies. Landscape plantings that incorporate trees and native plants also help cleanse the air stream and reduce dust and pollutants, capture nutrients in rainwater, and promote biodiversity by providing food and habitat to a host of birds, insects, and small mammals. Where water scarcity is a problem, the use of locally adapted native plants plays a particularly important role in the garden landscape, as does *xeriscaping*, a type of landscape design that primarily uses plants that have minimal water requirements.

The cultural use of cut flowers merits particular attention. Weddings, funerals, holiday celebrations, and a host of other traditions, including the American high school prom, all involve the use of cut flowers or foliage in the form of wreaths, floral arrangements, bouquets, or corsages.

Several key factors determine a plant's suitability for floral arrangements. It must have attractive flowers or leaves worthy of display. It must also have a suitable "shelf life." While some flowers look beautiful in the wild, they quickly droop and wilt when cut. A flower must be able to remain attractive for several days to be of practical use in floral designs. Some of the most popular cut flowers include roses (*Rosa* species), chrysanthemums (*Chrysanthemum* species), carnations (*Dianthus caryophyllus*), daisies and sunflowers (many different species in the Asteraceae family, most notably those in the genus *Helianthus*), lilies (Liliaceae), and orchids (Orchidaceae). Of these, chrysanthemums and carnations tend to have the longest shelf life and can stay fresh-looking for several weeks. Other plants such as the delicate baby's breath (*Gypsophila paniculata*) and the Boston fern (*Nephrolepis exaltata*) are used as "fillers" to fill up the space between flowers in an arrangement. Some kinds of plants retain their integrity when dried: strawflowers (*Bracteantha bracteata*), statice (*Limonium* species), and ornamental grasses are frequently used in dried arrangements.

sunflower (*Helianthus* species)

Many factors affect a plant's shelf life including the temperature, the water pH, humidity, and the food available to the plant after cutting. A weak acid added to the water will discourage bacteria from forming at the site where the flower was cut so the plant will be able to take up water longer. Sugar added to the water acts as a food source in lieu of photosynthesis. Floral preservatives are usually a combination of a mild acid and plant food.

Cool temperatures, high humidity, and minimization of exposure to wind will also help extend the shelf life of flowers. In addition, cut flowers should be kept away from ripening fruit or vegetables which give off a natural gas called *ethylene*. Ethylene is a plant hormone that speeds up the fruit ripening process but can also accelerate rotting of flowers. In the exercise on plant shelf life at the end of this unit you will experiment with several of these factors to see how they affect the freshness of your plant cuttings.

Plant Dyes

Before the development of synthetic dyes around 1850, the source of all colors used in paints, cosmetics, food, and fabric were natural dyes from materials such as plants, clays, and minerals. Plants that could be used to create desirable colors became important items of trade and production: for instance, indigo (*Indigo tinctoris*) produces a deep, rich blue, and henna (*Lawsonia inermis*) is the reddish hair dye originally used by Greek and Egyptian women.

Dyes will wash out of fabric or fade unless they are fixed by a compound called a *mordant*. A mordant chemically bonds the dye color to the fiber. In ancient times, people used wood ashes, rusty water, salt, vinegar, and pine needles as mordants. It is now known that salts of metals such as tin, copper, aluminum, iron, and chromium are the agents that cause colors to bond, or become color-fast. Alum is the most commonly used mordant today and can be purchased at drug and craft stores. **Caution:** Mordants are poisonous and should be handled with care and kept away from small children.

Natural fibers such as wool, cotton, silk, and linen take up natural colors best. Of these, wool seems to take up and display the colors more intensely. In one exercise in this unit, you will test your plant for its ability to produce a dye and its effect on different fabrics (as Luis did in his project, shown at left). Below is a list of some of the plants that people have used to create dyes of different colors:

Color	Plant source (common name)	Scientific name [family]
Yellow	Agrimony	*Agrimonia eupatoria* [Rosaceae]
	Dandelion	*Taraxacum officinale* [Asteraceae]
	Goldenrod	*Solidago* species [Compositae]
	Mullein	*Verbascum* species [Scrophulariaceae]
	Yellow onion	*Allium cepa* [Liliaceae]
	Tansy flowers	*Tanacetum vulgare* [Compositae]
	Spinach leaves	*Spinacia oleracea* [Chenopodiaceae]
Blue	Blueberries	*Vaccinium* species [Ericaceae]
	Hyssop leaves	*Hyssopus officinalus* [Labiatae]
Red	Cranberry	*Vaccinium oxycoccos* [Ericaceae]
	Beets	*Beta vulgaris* [Chenopodiaceae]
	Staghorn sumac fruits	*Rhus hirta* [Anacardiaceae]
Orange	Comfrey leaves	*Symphytum officinale* [Boraginaceae]
Green	Hyssop	*Hyssopus officinalus* [Labiatae]
	Lily-of-the-valley	*Convallaria majalis* [Convallariaceae]
	Parsley	*Petroselinum crispum* [Umbelliferae]
Violet	Elderberry fruits	
	Red cabbage	*Brassica oleracea* [Brassicaceae]
Brown	Red onion	*Allium cepa* [Liliaceae]
	Butternut hulls	*Juglans cinerea* [Juglandaceae]
	Walnut hulls	*Juglans nigra* [Juglandaceae]

A final note on plant dyes: Sources of natural dyes are regaining importance as people become more aware of the dangers inherent in synthetic dyes, especially those used in food and cosmetics. Would you rather eat a grilled cheese sandwich with cheese that gets its bright orange color from the chemical Red Dye #2 or from natural annatto (*Bixa orellana*)?

Plants for Perfumes and Fragrances

Cultures throughout the world from earliest recorded history have made use of plant fragrances. Ancient Egyptians were skilled perfumers by 3000 B.C. They burned frankincense (*Boswellia sacra*) and myrrh (*Commiphora* species) as incense in religious ceremonies. Their dead were embalmed with sweet-smelling herbs such as cedarwood (*Calocedrus decurrens*), cinnamon (*Cinnamomum verum*), and thyme (*Thymus vulgaris*). In ancient India, people made perfumes from sandalwood (*Santalum album*), jasmine (*Jasminum* species), and rose (*Rosa* species). Ancient Hebrews made ceremonial use of frankincense in their temples. The Chinese used jasmine in their religious rituals. The Greeks attributed the origin of plant fragrances to the gods. Some of their favored fragrances were mint (*Mentha* species), marjoram (*Origanum majorana*), thyme, rose, and hyacinth (*Hyacinthus orientalis*). The use of perfumes reached their high point in the Roman Empire; the Romans were lavish in their use of fragrant oils, adding them to clothing, bedding, wine, military flags, and their public baths.

Today we continue the ancient art of using plants for their fragrances. Following Cleopatra's example (she lured Antony not by looks alone), we employ hundreds of plant fragrances in perfumes, shampoos, deodorants, aftershave lotions, moisturizers, powders, and soaps so we will smell pleasing to others. Fragrances are subtle yet powerful in their ability to uplift people, set a mood, inspire memories, and attract romantic partners.

Aromatherapy is a popular practice that uses essential oils from plants to promote physical and mental well-being. Although little evidence exists to support claims for healing from major ailments or improvement of the immune system through aromatherapy, some research does show that the use of particular essential oils may have helpful effects on the body. In one study, the smell of nutmeg (*Myristica fragrans*) was found to sometimes reduce the blood pressure of people in stressful situations. One of the problems with assessing the validity of aromatherapy, however, is that the information about it often comes from commercial sources rather than from scientific laboratories that perform thorough, controlled experiments. As with all areas of medicine, we need to be able to distinguish marketing claims from solid scientific research.

In the following Activities section you will find experiments designed to help you test your plant for some of the uses presented here. As you conduct these experiments, you yourself may come up with a few new ideas for experiments — forge ahead!

Unit 6 Questions for Thought

On a separate piece of paper, answer the following questions as thoroughly as possible.

1. In addition to recycling, how could we produce paper without destroying our forests?

2. What properties in a plant fiber would make it a good source of material for paper?

3. What are some drawbacks to using synthetic pesticides?

4. If you were looking for a plant to use as a pesticide, what signs would you look for that would indicate that it might be a good one?

5. What is the purpose of companion planting?

6. Is the use of compost, green manure, or mulch the easiest method of enriching the soil? Why?

7. Is the use of compost, green manure, or mulch the most effective method of enriching the soil? Why?

8. If you wanted to create an organic fertilizer made from plant materials that contained 3 % nitrogen, 5 % phosphorous, and 2 % potassium, what plant materials would you use and in what ratios?

9. Describe how pH affects plant growth.

10. What are some safety recommendations for feeding wild animals?

11. Which kind of plant would be good for each of the following uses::
    a. houseplant
    b. landscape plant
    c. cut flower

12. Describe some methods for extending the life of cut flowers and other fresh plant materials such as fruits and vegetables.

13. What could you do to stop the color from fading in a dye?

14. Can you think of a good experiment to see if a particular plant fragrance could be used in aromatherapy? For example, if you heard that a plant fragrance calms people down, how would you prove or disprove this idea?

Unit 6 Activities

Can Your Plant Be Used to Make Paper?

Name: ______________________________ Date: ________________________

In this activity, you will make three types of paper and test the relative strength of each one. In Part A of the experiment you will prepare the following: 1) recycled paper from newsprint; 2) paper from your study plant; and 3) paper from a mixture of newsprint and your plant. In Part B of the experiment you will compare the strength of each sheet to a normal sheet of paper or newspaper.

A. Making Paper

Materials needed

Newspaper
Stems and leaves from your plant, dried and ground up
A square of window screen about 15 cm long on each side (screen from an art store designed for paper-making or a paper-making kit will produce better results)
A flat pan slightly larger than the screen
A blender, or a bowl and mixer
A rolling pin
One square of blotter paper (or paper towel), about 15 cm long on each side, for each sheet of paper to be made
Two bowls
Three plates (large enough to hold the sheets of paper)
Water

Directions

1. Cut or tear some of the newspaper into small squares.

2. Put the chopped newspaper into a blender (or bowl with a mixer), add just enough water and blend to create a mixture with a pasty consistency like that of a thick milkshake. Put this mixture into a bowl.

3. Put your dried, ground plant material into the blender, add water and blend to create a mixture with a thick, pasty consistency. Put this mixture into another bowl.

4. Place the screen in the bottom of the pan and cover it equally with the wet newspaper pulp. Holding the screen level, lift it up so the water drains out and the pulp remains on the screen.

5. Gently place the pulp-covered screen down on a flat surface covered in newspaper. Place the blotting paper down on top of the wet pulp.

6. Use the rolling pin to press out the excess water from the pulp.

7. Gently peel the blotting paper from the pressed pulp. Place the screen and pressed pulp on a plate with the screen on top.

8. Carefully lift the screen off the sheet of pressed pulp. (Try not to let it rip.) Let it dry on the plate.

9. Rinse out the screen and pan.

10. Repeat steps 4 through 9 using the plant mixture instead of newspaper to make paper.

11. Repeat steps 4 through 9 using a mixture of half newspaper pulp and half plant pulp to make paper.

B. Testing Paper Strength

Materials needed

A normal sheet of non-recycled paper (you may also include other types of store-bought papers such as newsprint, recycled paper, and/or that made from different plant materials like hemp or banana plant fibers.)
The dried sheet of paper made from your plant
The dried sheet of paper made from newspaper pulp
The dried sheet of paper made from a mixture of your plant and newspaper pulp
Scissors
Hole punch
Spring scale
Graph paper and pencil

Directions

1. Cut each of the papers into equal-size squares. (Try to make four 2 × 2 inch squares from each sheet.)

2. Label equal square so you know what type of paper it is.

3. Punch a hole in the middle of each square.

4. Place the hook of the spring scale into the hole in the first square of paper. While holding the square of paper attached to hook, gently pull on the other end of the spring scale. Watch carefully to see how much force is needed to rip the paper. In the chart below, record the type of paper tested and the force needed to tear it.

5. Repeat this procedure for each square of paper.

6. Graph your results to see how paper type relates to average amount of force needed to tear the paper.

Number of samples	Type of paper	Force needed to tear	Average force needed to tear
________	________	________	________
________	________	________	________
________	________	________	________
________	________	________	________
________	________	________	________

Analysis

1. Which of the three types of papers that you made was the strongest? The weakest?

____

2. Based on this experiment, do you think your plant is a potential source of fibers for making paper? Why or why not? ________________________

____

3. Do you think your plant could be mixed with other plant fibers or with recycled paper to make a strong, attractive paper? ________________________

____

4. What do you think are the most important characteristics are for a plant to be a good source of paper? ________________________

____

Can Your Plant Be Used for Cordage or Construction?

Name: ______________________ Date: ______________________

Perhaps your plant has fibers than can be used to make cordage (rope or twine), or plant parts that can be used to construct or weave a useful object. In this exercise, you will develop your own hypothesis about the potential usefulness of your plant to make something other than paper. You will design an experiment to test your hypothesis. Use your imagination. Here are two suggestions:

A. Try braiding, twisting or weaving your plant fibers into rope or twine. (Sometimes fibers are more flexible and less liable to break if you soak them prior to working with them.) Is it possible to weave the fibers together to make a mat or basket?

B. Construct a box by gluing together dried stems from your plant. When dry, test the strength of the box using weights. How much weight will the box hold before it collapses?

Analysis

1. What do you want to find out from this experiment? What is your hypothesis?

____

____

____

____

2. Describe how you set up your experiment.

____

____

____

____

____

____

____

____

3. How did you measure the outcome of your experiment?

____

____

____

____

4. Based on these observations, does your plant produce fibers that are useful for something other than paper-making?

____

____

____

Can Your Plant Be Used to Make an Insecticide?

Name: ______________________________ Date: ________________________

In this exercise, you will design an experiment to see if you plant might contain chemicals that kill or repel insect pests. Meal worms or grasshoppers, both available at pet stores, are just two examples of appropriate specimens to work with; another would be the abundant and destructive Japanese beetle. (**Note:** Most states permit experiments that may cause mortality to invertebrates such as insects. However, you should check local regulations regarding this prior to conducting such an experiment. Do not use rare or beneficial insects in these experiments.) Below are four suggestions for conducting this experiment.

A. Prepare an extract from your plant and put it in a spray bottle. You can test this potential insect spray on houseplants that are infested with scale, mealy bugs, or aphids. You could compare the effect of your plant spray with a plant that is treated with only water (your control) and a plant that is sprayed with a commercial insecticide formulated especially for houseplants. (If you have only one infested houseplant you could remove three or most branches or leaves that have approximately the same amount of insect pests. Conduct your experiment on these three plant sections, choosing one as the control.

B. Test your plant extract on insects in enclosed chambers to see if it kills or repels them. Use at least two clean plastic soda bottles converted into chambers and place an equal number of insects in each one. Pin holes will provide adequate oxygen for the insects. Place a cotton ball soaked in your plant extract in one bottle and a cotton ball soaked in water (your control) in the other. Observe whether your extract repels or harms the insects. Alternatively, you could make a maze (plastic tubes and soda bottles work well) to see if the insects react to your plant spray by moving towards it, away from it, or show no reaction at all.

C. To control an insect pest, one needs to understand its life cycle. For example, some treatments will affect an insect in a larval stage but not in the adult stage, or vice versa. You might want to observe how effective your plant spray is on an insect on the various stages of its life cycle. Use entomology books or insect field guides to help you understand and identify these stages.

D. You may want to see if your plant spray is effective at deterring other animals that eat garden plants such as rabbits.

Analysis

1. What do you want to find out from this experiment? What is your hypothesis?

____

____

____

____

2. Describe how you set up this experiment.

____

____

____

____

3. What is the name of the insect you are using in this experiment? What stage in its life cycle are you using to test your plant extract?

4. How did you measure the outcome of the experiment?

5. Based on this experiment, what do you think of your plant's potential use as an insecticide?

Does Your Plant Extract Stimulate Growth in Other Plants?

Name: ______________________________ Date: ________________________

In this exercise, you will conduct an experiment to see if your plant extract stimulates the growth of another type of plant. This will help you assess the potential of your plant as a growth stimulant or fertilizer.

Materials needed

Numerous fast-growing seedlings such as tomato or corn that have been raised under similar conditions and that are similar in their height, size, and vigor
Three large pots of the same size each with labels
Potting soil
Your plant extract
A one-liter jar
A sunny window or place to raise the seedlings
Commercial fertilizer
Metric ruler

Directions

1. Transplant equal numbers of seedlings (5 to 10) into each pot and cover the roots with potting soil.

2. Label each pot with your name, the start date, and the treatment for that pot as follows: (a) Control (no treatment); (b) Fertilizer; and (c) Plant Extract.

3. Water the pot labeled Control with one-half liter untreated water.

4. Water the pot labeled Fertilizer with one-half liter water and the appropriate amount of fertilizer as specified on the box. (Too much fertilizer will harm the plants. Follow the recommended dosage exactly.)

5. Water the pot labeled Plant Extract with 450 ml of water mixed with 50 ml of your plant extract.

6. Place all three pots near a sunny window where you can observe them. Water them with plain, untreated water whenever the soil begins to look and feel dry. (They do not need fertilizer again for several months.) Be sure to give all three plants equal amounts of water at the same time for each watering.

7. After one month, evaluate the effect of the different treatments. You may cut the plants at the base of the stem and weigh the plants from each pot to compare any differences in weight. Or, if you prefer not to kill your plants, you can count the number of leaves and measure stem circumference at the plant base, leaf width, and plant height.

Analysis

1. Using the following chart, describe the what happened to each pot of seedlings.

Pot group	Date	Height (avg.)	Vigor/health	Color	Disease/pests	Other observations
________	__/__	________	________	______	________	________
________	__/__	________	________	______	________	________
________	__/__	________	________	______	________	________

2. Can you think of another method to determine whether your plant would make a good fertilizer or growth stimulant?

3. Based on the results from this experiment, do you think your plant extract stimulates growth in other types of plants? Does it inhibit growth?

What Is the pH of Your Plant Extract?

Name: ______________________________ Date: ________________________

In this exercise you will measure the pH of your plant extract using pH indicator paper (litmus paper). You may also use a pH meter if one is available.

Materials needed

Litmus paper or a pH meter
Your plant extract
Distilled water
Clean test tubes

Directions

1. Dip the litmus paper or pH meter probe in a test tube with your plant extract. If you are using litmus paper, compare the color it changes to with the color chart that comes with the paper. The chart will tell you what pH level each color represents. If you are using a pH meter, follow the manufacturer's instructions for using the meter and reading the pH level. Record the pH in the chart provided below.

2. Dip the litmus paper or pH meter probe in a test tube with distilled water. Identify the pH level as described in Step 1. Record the pH of the distilled water below.

3. For comparative purposes, record the results of some or all of the pH measurements your classmates found for their plants.

Plant name	pH of extract
____________________	________
____________________	________
____________________	________
____________________	________
____________________	________
____________________	________
____________________	________
____________________	________
____________________	________
____________________	________
____________________	________
____________________	________
Distilled water	________

Analysis

1. How do you think your soil will be affected by the addition of your plant extract: Will it become more acidic or more basic? ______________________________

____

2. Would the pH of your plant extract be helpful or harmful to garden plants? Why?

____

____

Optional activities

A. Measure the pH of the soil your where your plant is growing. Collect a sample from the upper 2 or 3 inches of soil where your plant seems to be particularly thriving Put 10 grams of soil in a test tube with 20 milliliters of distilled water. Shake and let the soil settle out. Measure the pH of the water in the test tube. Be sure to test at last two samples. Under what soil pH conditions does your plant appear to grow well?

B. Some scientific companies sell kits that enable you to analyze plant tissue for the presence of specific minerals, most commonly nitrogen, phosphorus, and potassium. (Some companies that sell kits are listed in Appendix A.) Knowing the quantity of mineral nutrients held in plant tissue is useful for a variety of reasons. You can estimate how much "fertilizer" a plant would contribute to the ecosystem if it were incorporated into the soil as mulch or compost. You could also make mineral-uptake comparisons between different types of plants grown in the same soil, or between the same type of plant grown in soils of different fertility levels. Knowing the quantity of nitrogen, phosphorus, and potassium in the tissues of your study plant may shed some light on its effectiveness in stimulating plant growth in the previous lab activity.

Could Your Plant Be a Source of Feed for Animals?

Name: ________________________ Date: ____________________

In this exercise you can design your own experiment to test the potential use of your plant as a feed for a particular type of animal. Before conducting any tests, however, be sure your plant will not cause any toxic effects to the animal(s) you are using in your experiment. Below are two suggestions for experimental design.

A. Set out your plant along with several other food choices so the animal being tested can select the food it prefers. After several trials you should have a good idea whether or not the food is palatable to that animal.

B. Using bird feeders, you can test your plant's potential as a food for wild birds. Set out different kinds of foods in several different bird feeders, including such things as sunflower or pumpkin seeds, berries, millet and other grains, other bird seed, and your plant. Make observations over a period of several weeks about bird species and specific food choices. You may also want to try setting out different parts of your study plant (leaves, flowers, roots, seeds, etc.) as feed to see if any particular part of the plant is preferred.

Analysis

1. What is it that you want to find out from this experiment? What is your hypothesis?

____

____

____

2. Describe how you set up your experiment. ____________________

____

____

____

____

____

____

3. What type of animal(s) are used in your experiment? What does their normal diet consist of? ____________________

____

4. How did you measure the outcome of this experiment?

____

____

____

5. Based on this experiment, how well does your plant rate as a food source for the type of animal(s) you studied? ____________________

____

____

Could Your Plant Be Used in Fresh Floral Arrangements?

Name: ______________________________ Date: ________________________

In this experiment you will test your plant to see if it would make a good cut flower or foliage plant for flower arrangements. You will be checking to see how long a shelf life it has under various conditions after being cut. (i.e. Cut plants will be subjected to five different treatments; each treatment will be replicated and stored under two different temperature regimes). You will compare the responses of your plant to these conditions with those of a commercial cut flower such as a carnation.

After conducting this experiment, you will be able to answer the following relevant questions: (a) Does sugar extend the shelf life of your plant after cutting? (b) Does vinegar discourage bacteria from growing on your plant's cut stem? (c) How does temperature affect the shelf life of your plant? (d) Do floral preservatives extend the shelf life of your plant?

Materials needed

Approximately 10 live plants (or large stems) of your plant species
Scissors or sharp knife
Ten plastic cups, vases, or containers
Teaspoon
Water
Sugar
Vinegar
Floral preservative (optional)
Ten carnations or other popular cut flower
Access to a refrigerator
Marker or labels for the containers

Directions

1. Label the 10 containers as follows:

A1. Plain water, refrigerated
B1. 2 tsp sugar, refrigerated
C1. 1 tsp vinegar, refrigerated
D1. 1 tsp sugar, 1 tsp vinegar, refrigerated
E1. 2 tsp floral preservative, refrigerated

A2. Plain water, room temperature
B2. 2 tsp sugar, room temperature
C2. 1 tsp vinegar, room temperature
D2. 1 tsp sugar, 1 tsp vinegar, room temperature
E2. 2 tsp floral preservative, room temperature

2. Add one cup of water to each cup. Add the treatments (sugar, vinegar, sugar & vinegar mix, and floral preservative) exactly as specified on the labels.

3. Cut the flower stems or foliage that your are testing from your live plants with the knife or scissors. Make sure each one is approximately the same size and that all cuttings are fresh and healthy. Place one cutting in each of the ten labeled containers. (If you have enough cuttings to put more than one stem into each container, you may do so. Make sure each container holds the same number of cuttings.)

4. Put one carnation into each of the containers. You will be comparing the shelf life of the carnation to that of your plant.

5. Put containers A1, B1, C1, D1, and E1 into the refrigerator. Put the other containers on a table in the room.

6. Make observations every two days for two weeks, recording the condition of the cuttings on the data sheet below.

Use the following numbers to describe the health of your cuttings:

1 = dead
2 = badly wilted or dried out
3 = slightly wilted
4 = slightly limp but still fresh -looking
5 = fresh and healthy-looking

A. Data on the shelf life of my plant

Treatment	Number of days after cutting 2	4	6	8	10	12	14
A1. Water, refrigerated	__	__	__	__	__	__	__
A2. Water, room temp	__	__	__	__	__	__	__
B1. Sugar, refrigerated	__	__	__	__	__	__	__
B2. Sugar, room temp	__	__	__	__	__	__	__
C1. Vinegar, refrigerated	__	__	__	__	__	__	__
C2. Vinegar, room temp	__	__	__	__	__	__	__
D1. Sugar & vinegar, refrigerated	__	__	__	__	__	__	__
D2. Sugar & vinegar, room temp	__	__	__	__	__	__	__
E1. Floral preserv., refrigerated	__	__	__	__	__	__	__
E2. Floral preserv., room temp	__	__	__	__	__	__	__

B. Data on the shelf life of carnations

Treatment	Number of days after cutting 2	4	6	8	10	12	14
A1. Water, refrigerated	__	__	__	__	__	__	__
A2. Water, room temp	__	__	__	__	__	__	__
B1. Sugar, refrigerated	__	__	__	__	__	__	__
B2. Sugar, room temp	__	__	__	__	__	__	__
C1. Vinegar, refrigerated	__	__	__	__	__	__	__
C2. Vinegar, room temp	__	__	__	__	__	__	__
D1. Sugar & vinegar, refrigerated	__	__	__	__	__	__	__
D2. Sugar & vinegar, room temp	__	__	__	__	__	__	__
E1. Floral preserv., refrigerated	__	__	__	__	__	__	__
E2. Floral preserv., room temp	__	__	__	__	__	__	__

Analysis

1. Which treatment extended the shelf life of your plant cutting the most?

____

2. Which treatment extended the shelf life of the carnations the most?

____

3. Which treatment(s) allowed the cuttings and carnations to wilt the fastest?

____

4. How did refrigeration affect the shelf life of the different groups of flowers?

___

5. How long did your cutting stay fresh in water at room temperature? How did this compare to the carnation in water at room temperature for the same length of time?

___

___

___

6. Based on this experiment, do you think your plant has a long enough shelf life to be marketed as a cut flower or foliage plant? ___

___

Optional activities

A. How does ethylene gas affect the shelf life of your plant? This addition to the above experiment will allow you to observe the effect of ethylene gas, a natural plant by-product, on your plant and on the carnation. Put a plant cutting and a carnation in a cup of plain water. Carefully place the cup into a plastic bag with a banana and seal it. (Ripening fruits like this banana give off ethylene.) Observe how the cutting and carnation are affected over the next 14 days, compared to the ones not exposed to ethylene.

B. Does your plant add beauty to floral arrangements? Create a floral arrangement using cuttings from your plant. Use your plant in different ways to add texture, color, and shape to various arrangements of flowers. Ask your classmates for their opinions and ideas.

C. Does your plant show potential as an attractive houseplant? Design a way to test the potential success of your plant as a houseplant. Can it survive the reduced light levels and continual warmth associated with growth indoors?

D. Does your plant show potential as a landscaping component? Learn about the various criteria necessary for landscaping plants in your area. Can it survive local climatic conditions? How would it contribute to an overall landscape design? (color, texture, height, use in border plantings, a source of food for wildlife?) Design your own test to determine your plant's potential in as an ornamental.

Can Your Plant Make a Colorful Dye?

Name: ______________________________ Date: ________________________

In this experiment, you will make a dye from your plant and see if it creates an attractive color on various kinds of fabrics.

Materials needed

Hot plate or access to a stove
Strips of different types of white fabric including cotton, wool, and a synthetic fabric like polyester (you may substitute or include wool yarn for the fabric)
1.15 kg of your plant material. You can use flowers, roots, leaves, or fruits of your plant, depending on what has the most desirable color. You may want to try making different dyes from different plant parts
⅙ cup alum or ¼ cup white vinegar
3-quart cooking pot (**not** an aluminum pot)
Clothes hangers or a rack to dry the fabric on
Water
Newspaper
Slotted metal stirring spoon

Directions

1. Fill the pot with two quarts of water and add the alum or white vinegar.
2. Bring the mixture to a boil and add the plant material.
3. Turn the heat down and let it simmer for 30 minutes.
4. Wet the strips of fabric in cool water.
5. Turn the heat off and remove the plant material with the slotted spoon.
6. Add the pieces of wool, cotton and synthetic fabric to the pot and stir.
7. Let the material sit the dye for at least 15 minutes and up to 24 hours.
8. Remove the fabric strips and hang to dry with newspaper underneath to catch the drips.

Analysis

1. What part or parts of the plant did you use to make the dye(s)?

____

____

2. How would you describe the color your plant produced? Do you find it attractive?

____

____

3. How did the different fabrics respond to the dye? Which material has most intense color? The least intense color? ______________________________

____

____

____

4. Do your think your plant produces a good dye? ______________________

____

Does Your Plant Produce a Pleasant-Smelling Fragrance?

Name: ______________________________ Date: ________________________

In this exercise you will conduct a survey to find out how people react to the fragrance of your plant. You will compare people's reaction to four different plant-based fragrances.

Materials

Your plant extract
Three other plant-based fragrances (such as vanilla, cinnamon, rose oil, clove, lemon, thyme)
Four bottles with caps
Aluminum foil
10 copies of the survey form below

Directions

1. Find 10 people willing to participate in your survey and make 10 copies of the survey form that appears on the following two pages.

2. Cover your bottles with aluminum foil so the contents will not be visible to the people being surveyed.

3. Label your bottles A, B, C, and D.

4. Put one fragrance in each bottle. Write down the name of each plant fragrance and the bottle in which it was placed.

5. Give each participant a copy of the survey form. Ask them to smell one bottle at a time and fill out the survey sheet for that bottle before moving on to the next one.

Analysis

1. What four fragrances did you use in your survey?

A. ___________________________ C. ___________________________

B. ___________________________ D. ___________________________

2. How did people rate the four fragrances?

A: Total: _____ Average rating _____ Answers ranged from ___ to ___

B: Total: _____ Average rating _____ Answers ranged from ___ to ___

C: Total: _____ Average rating _____ Answers ranged from ___ to ___

D: Total: _____ Average rating _____ Answers ranged from ___ to ___

3. Which plant extract did most participants prefer? ___________________________

4. What words did most people choose to describe your plant extract?

____

____

5. What feelings did most people associate with your plant extract?

____

____

6. Based on this experiment, do you think your plant shows potential for development as a perfume or fragrance? ____

____

Plant Fragrance Survey Form

Name of person surveyed ______________________________ Date ______________

Plant Fragrance A:

1. On a scale of 1 to 10, how pleasant do you find this fragrance?

intolerable so-so the best

1 2 3 4 5 6 7 8 9 10

2. Circle the word that best describes this smell:

fresh	sweet	sour	pungent	bitter	disgusting
hot	spicy	refreshing	minty	foul	can't describe

3. What feelings do you associate with this fragrance?

sleepy	irritable	happy	refreshed	angry	thoughtful
alert	peaceful	sad	stressed	loving	energized
sick	rejuvenated	spiritual	bored	other	____________________

4. What plant do you think this fragrance comes from? ______________________________

Plant Fragrance B:

1. On a scale of 1 to 10, how pleasant do you find this fragrance?

intolerable so-so the best

1 2 3 4 5 6 7 8 9 10

2. Circle the word that best describes this smell:

fresh	sweet	sour	pungent	bitter	disgusting
hot	spicy	refreshing	minty	foul	can't describe

3. What feelings do you associate with this fragrance?

sleepy	irritable	happy	refreshed	angry	thoughtful
alert	peaceful	sad	stressed	loving	energized
sick	rejuvenated	spiritual	bored	other	____________________

4. What plant do you think this fragrance comes from? ______________________________

Plant Fragrance C:

1. On a scale of 1 to 10, how pleasant do you find this fragrance?

intolerable so-so the best

1 2 3 4 5 6 7 8 9 10

2. Circle the word that best describes this smell:

fresh	sweet	sour	pungent	bitter	disgusting
hot	spicy	refreshing	minty	foul	can't describe

3. What feelings do you associate with this fragrance?

sleepy	irritable	happy	refreshed	angry	thoughtful
alert	peaceful	sad	stressed	loving	energized
sick	rejuvenated	spiritual	bored	other ______________	

4. What plant do you think this fragrance comes from? ______________

Plant Fragrance D:

1. On a scale of 1 to 10, how pleasant do you find this fragrance?

intolerable so-so the best

1 2 3 4 5 6 7 8 9 10

2. Circle the word that best describes this smell:

fresh	sweet	sour	pungent	bitter	disgusting
hot	spicy	refreshing	minty	foul	can't describe

3. What feelings do you associate with this fragrance?

sleepy	irritable	happy	refreshed	angry	thoughtful
alert	peaceful	sad	stressed	loving	energized
sick	rejuvenated	spiritual	bored	other ______________	

4. What plant do you think this fragrance comes from? ______________

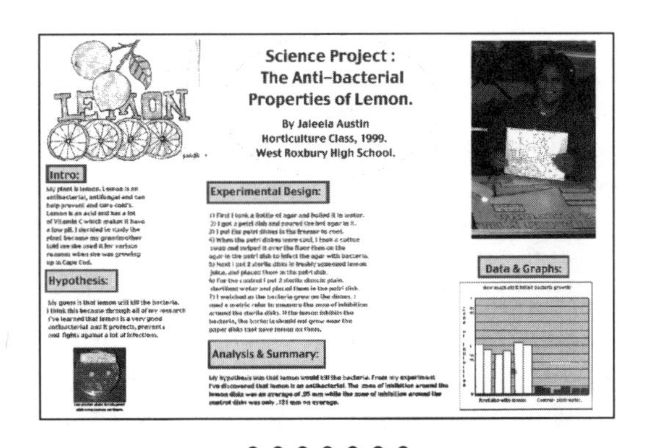

Unit 7
Designing Your Own Experiment

Introduction

The objective of this unit is to help you design and conduct your own unique experiment about your plant. Now is your chance to pursue a question you may have been wondering about. My students have investigated many questions such as the following: Can I make dye from blueberries? Can tea tree oil clear up acne? Can chamomile tea bring out the highlights in hair? Can the oil of eucalyptus leaves repel insects?

No magic is required to create your own experiment. The key lies in asking a specific question that you can test for. Once you have your question, you come up with the steps to find an answer. Your curiosity and imagination will help you develop your experiment. Your care in following the "recipe," the experimental procedure, will help you succeed.

In the first part of this unit we emphasize safety. We discuss FDA regulations associated with the development of new pharmaceutical drugs and herbal supplements as an introduction to the issues of safety and ethics in experimental design. We follow that with guidelines on safety precautions for working with plants in developing your ethnobotany experiment. Read these guidelines carefully before embarking on any experiment. No matter what plant you choose for your experiment, you need to research the plant's safety to make sure it can in no way cause harm to you or others.

In the second part of this unit you will find suggestions for choosing a research question and developing your methods. Guidelines for recording your observations, and analyzing your data using graphs and simple statistics are also included. Finally, we make a few suggestions for developing the final presentation of your results.

Throughout this unit I have included examples of actual projects in ethnobotany created by ninth-, tenth-, and eleventh-grade students at West Roxbury High School in Massachusetts (such as Jalela's poster about lemon, above). These examples, taken from these students' own experiments, will clarify the information presented here and stimulate some ideas of your own. If you follow these procedures carefully and use imagination in planning your experiments, you can make exciting new discoveries.

The Importance of Safety and Ethics in Experimental Design

FDA Regulations and the Development of New Drugs and Herbal Supplements

The United States Food and Drug Administration has very stringent regulations concerning research protocol for the development of pharmaceutical drugs. Scientists work for many years conducting laboratory research on new compounds, to identify their active ingredients and understand how they work, before they are allowed to test these drugs on people in clinical trials. It would be unethical to test any compound on people before being certain that it safe to use. This extensive period of lab testing is one reason that the development of new pharmaceuticals takes so long and is so expensive, pushing the development costs of each new drug into many millions of dollars.

Federal regulation of herbs in the United States differs from that of pharmaceutical drugs. Less testing is required to bring a new herb to market than is required for a new drug, as long as the herb is marketed as a "dietary supplement" and not as a drug or medicine. Although the herbal supplement must be proven safe for consumption, their modes of action and active ingredients are not assessed in the same way that the FDA requires for pharmaceuticals. According to the 1994 U.S. Dietary Supplement Health and Education Act, those who develop and market herbal medicines cannot claim that their product cures a specific illness. They may only claim that their product promotes a general well-being of the body or that it promotes a bodily function.

For example, a bottle of pills made from purple coneflower, *Echinacea purpurea*, may say something like "This product nutritionally supports healthy immune function"; however, it may not say that echinacea prevents colds or flu. A label on a bottle of herbal extract of

Beyond Reading the Label

One of the words that marketers use to appeal to consumers is "natural." If you read the labels of the herbal supplements you find in your local drug store or health food store, you will see this and similar terms. Poison hemlock is natural, as is the *E. coli* bacterium — yet both are also very dangerous and potentially deadly. Use of the word "natural" is no guarantee that a product is also safe. Manufacturers of supplements are responsible for their own premarket testing and the truthfulness of the claims on the labels, so consumers need to be cautious and, above all, well-informed. Here are some guidelines for choosing dietary supplements:

- Look for products with the "U.S.P." notation. This indicates that the manufacturer has followed rigorous "United States Pharmacopeia" standards.
- Be skeptical of any product that is touted as a "breakthrough" or "new discovery," as having "magical" properties, or as being an "ancient secret."
- Don't be fooled by pseudo-medical jargon such as "detoxify," "purify," "rejuvenate," and "energize."
- Question any product that claims to treat a wide variety of symptoms.
- Be suspicious of products with no side effects, only benefits.

bilberry (*Vaccinium myrtillus*; a small, wild cousin of the blueberry) may say something like "This product nutritionally supports visual function"; it may not say that bilberry improves eyesight or corrects vision problems. In addition, all such bottles must include statements that accompany their general claims to improve a bodily function with a disclaimer that may read as follows: "This statement has not been evaluated by the Food and Drug Administration. This product is not intended to diagnose, treat, cure, or prevent any disease."

This difference in regulations between dietary supplements and pharmaceuticals means that the development of herbal products requires less research time and expense to receive FDA approval. On the positive side, this makes it easier and faster to bring new products to market. On the negative side, less may be known about the biochemistry and function of many "dietary supplements" being brought to market; hence, the herbal market may be more susceptible to misinformation and fraudulent claims. Nonetheless, according to government regulations, both dietary supplements and commercial pharmaceuticals must be proven safe for use by people in appropriate dosages before they reach the market.

In summary, to develop new drugs and herbal supplements for the U.S. market, scientists plan, carry out, and replicate an extensive number of tests to learn more about how these products work and to prove them safe before bringing them to market. Each experiment yields a limited but useful piece of information that contributes to the overall understanding and analysis of the plant or compound under study. The testing procedures are required to meet government standards of safe and ethical experimentation.

In this unit, you will learn to approach your research the same way scientists do when they investigate new medicinal compounds for the first time. For plants that are not common foodstuffs, you must develop experiments that do not involve testing your plant directly on people or animals. Tests must be ethical and not cause pain or discomfort to either people or animals. Guidelines for the design of safe experiments are outlined in the next section; read them thoroughly so that you become familiar with them.

Commission E? What's That?

Although many herbs have not had rigorous research conducted on them in the United States, some countries — notably Germany, Japan, China, and India — have conducted intensive clinical trials to study the medicinal properties of many herbals products.

In Germany, for example, herbal medicines are often taken with a doctor's prescription and paid for by national health insurance. Over the past 20 years, Germany's national health agency has published very detailed reports that analyze and summarize international clinical studies and toxicological assessments of hundreds of herbs to assess their safety. These reports are called the Commission E monographs. The American Botanical Council has translated the complete Commission E monographs into English.

A professor in the School of Pharmacy and Pharmaceutical Sciences at Purdue University called the Commission E reports "the most accurate information available in the entire world on the safety and efficacy of herbs and phytomedicines." Although the publication is very expensive, if you want to refer to the Commission E reports for your own research, you may be able to use them at the library of a nearby college or university or request a specific portion of them through interlibrary loan at your local library.

Safety Considerations

You must be sure your experiment is a safe one. On the next page is a box containing a list of some important safety precautions. You must familiarize yourself with them thoroughly before you begin designing or conducting experiments.

Your project must be in compliance with all local, state, and federal regulations and standards. Some laws regulate experimentation. For example, most places have specific guidelines about what constitutes an acceptable science fair project (e.g., rules about the use of animals in experiments). Other laws safeguard human and environmental health. For example, the disposal of chemical and biological wastes is regulated to protect our health and the health of our water, air, and soil. Likewise, quarantine laws regulate the transport of plants across state and country borders to guard against the introduction of potential pests and to stop the transport of illegal plant materials.

You or your teacher should also determine what regulations beyond the guidelines on the next page are in effect for your area so your experiment does not violate them. Consult the following sources for further information on safety issues in the United States:

Office of Protection for Research Risks
National Institute of Health
9000 Rockville Pike
Bethesda, MD 20205

Safety in the Secondary Science Classroom (stock #471-14652), available from:
National Science Teachers Association
1742 Connecticut Ave. NW
Washington, DC 20009

Safety in the High School, available from:
American Chemical Society
Career Publications
1155 16th St. NW
Washington, DC 20036

Read This First!

General Guidelines for a Safe Ethnobotanical Science Experiment

1. Do not use disease-causing pathogens (i.e., microbial agents that may cause illness including all potentially dangerous species of fungi, bacteria, viruses, or protozoa) in any experiment. You may use microorganisms that are harmless to people (e.g., the bacteria that produce yogurt, spores from yeast used to make bread) if they are handled and disposed of correctly.
2. Do not use microbial agents used in genetic engineering or recombinant DNA experiments.
3. Handle all sharp objects including razor blades, syringes, needles, knives, and so forth, with extreme care and only under a teacher's guidance. Dispose of any sharp objects in accordance with local regulations.
4. Do not use dangerous chemical substances in your experiments. These include compounds that are flammable, explosive, corrosive, poisonous, carcinogenic, or mutagenic. Check with your teacher concerning the use of any chemical compounds and their potential dangers. Use chemical compounds only under direct supervision.
5. Do not experiment with drugs, prescription medications, or over-the-counter medications (e.g., aspirin) in your experiments.
6. Do not experiment with poisonous or illegal plant materials.
7. Do not use plant materials that have entered the country illegally.
8. Do not use vertebrate animals (mammals, birds, reptiles, amphibians, or fish) in any experiment that might cause them harm or death.
9. You may use invertebrate organisms (e.g., insects, brine shrimp, protozoa) in an experiment if permitted by local and state regulations.
10. Do not ingest, inhale, or apply to the body any plant materials other than safe, legal foodstuffs or herbs that are commonly available in supermarkets, health food stores, or ethnic markets in the state where the research is being conducted.
11. Do not ingest plants collected from the wild unless assisted by a qualified botanist who is certain of their identification. Since many plants are poisonous, misidentification of a wild plant could prove fatal.
12. Do not use tobacco or alcoholic beverages in experiments involving humans or other vertebrate animals.
13. Never involve or expose people in any activity or test that might cause them physical, mental, or emotional stress, pain, or hardship.
14. Use clean, preferably sterile, tools in all your work, as well as any lab safety gear (such as goggles, gloves, and so on) required by your school. Refrigerate fresh plant parts, teas, decoctions, and any other perishable materials soon after preparation when not in use and discard them after they become old, rancid, or spoiled.
15. If your experiment involves eating any kind of food, make sure that the food to be consumed is fresh and thoroughly cleaned. Dirty or spoiled food can cause food poisoning and other severe illnesses. If the experiment involves meat, it should be well-cooked and tested with a meat thermometer according to the manufacturer's instructions. After preparation, food items should be used right away or refrigerated and served within 24 hours. Utensils and containers used in the preparation and storage of foodstuffs should be thoroughly clean before and after use.
16. Never prepare food in a science laboratory where chemicals are used. Food that will be consumed by humans should be prepared in a kitchen with kitchen tools following all health and safety precautions that normally go into preparing food.
17. Do not ingest spices in quantities of more than ¼ teaspoon, or a "pinch."

Planning Your Experiment

In science, a good experiment is rarely an end in itself. A well-designed experiment will yield an answer that leads to another question; well-designed experiments with outcomes that can be validated can sometimes lead to new products or to new policies or theories that influence society. But the scientific process is more akin to careful detective work than it is to digging for buried treasure. You follow a trail of clues, step by step, collecting evidence along the way. Sometimes you're sidetracked; sometimes you're frustrated, and sometimes you make surprising and exciting discoveries. An experiment can supply you with a piece of the puzzle, not the whole picture.

The Notebook

Prior to beginning your experiment you need to get a bound notebook where you will write down everything related to your research. All scientists use notebooks (or lab books) to record their observations, whether they are biochemists testing plant compounds to treat HIV and AIDS, anthropologists studying farming methods in Borneo, or high school students experimenting with sage (*Salvia officinalis*) as an insect repellant. Your notebook serves as a daily chronicle of what you do during each step of the experiment, and what happens as a result. You can record your experimental data in your notebook along with any informal observations. Use your notebook to record questions, ideas, and concerns as they arise during your work.

While conducting an experiment, unexpected things happen or problems arise. As a result, you may get a new idea or inspiration that leads your research in a new direction. It is crucial to keep a record of such events in your notebook. Later on, these notes may provide critical insights about the meaning of your research results.

The main rule to remember about your notebook is that each entry should be dated. Record your data and observations carefully and legibly. (Don't get caught later saying "Is that a 1 or a 7?") Don't try to commit any of your experiment to memory; even the most seemingly insignificant details should be recorded as you go along. Write down each part of your experiment in your notebook including your question, hypothesis, experimental design, and data. You can also use your notebook to record any drawings, diagrams, graphs, poetry, insights, and ideas related to your experiment.

Here are two sample entries from the notebook of Jamika (shown here writing):

> November 12, 1996: I set up two bowls side by side, one with lettuce, one with collard greens. The guinea pig went right to the lettuce 3 times. I expected the guinea pig to prefer the collards which have more vitamin C. Maybe he likes the crispness of the lettuce. Why does he prefer the lettuce when he needs Vitamin C?
>
> November 14, 1996: Today the guinea pig preferred the collard greens once and the second time went for the lettuce. Why did he like the collards today? Was it because the lettuce was more wilted today? If I stopped giving him Vitamin C for a few days would he want the collards more?

Reviewing the Literature

As part of a project in ethnobotany, your experiment should fit into the larger picture of the connection between your plant and its use by people. A thorough review of the literature will help you understand your plant in this context. As you read, be aware of the difference between, on the one hand, reports about traditional uses based solely on interviews and, on the other hand, actual results from scientific experiments. Understanding this difference will help you choose what to test for, and will help you see how your research can contribute to the current body of knowledge about your plant. After finding out what other experiments scientists have conducted, you may decide to repeat an experiment to see if you get similar results, or you may choose to try your own new experiment.

The directions for completing a research paper are detailed in Unit 2. If you have not already completed your background research, return to that section for information on how to do so.

Choosing What to Test For

Deciding on a focus for your research will depend on what you have learned during your review of the literature and from your informant. This information about your plant should lead to a question, the question will lead to an experiment, the experiment should produce results, and the results will (hopefully, but not always) answer your question. That answer is your goal.

As you consider research questions and potential experiments, bear in mind these two factors: (1) that your experiment is a safe and ethical one as outlined in the previous section; and (2) that your experiment is within your budget and the materials available to you.

My Contributions to Science

You don't have to have a Ph.D. to make a contribution to science. The work you are doing right now could contribute to the larger body of knowledge about your plant, in a few ways.

First, your research may involve conducting an experiment that has never been recorded in the literature for your plant. As such, your research would contribute new information to science about your plant.

Second, you might choose to repeat an experiment that has already been conducted and reported in the literature. The replication of experiments is an important part of the scientific process; it helps demonstrate how reliable the results are from the prior experiment.

Third, you may choose to repeat an experiment reported in literature using new methods that you develop yourself. Your improvement of experimental methods might produce more applicable or more accurate results than the other procedure.

The Question

You may have several questions about your plant that come to you while conducting your background research. List all your potential questions in your notebook. Later, you can reflect on this list and think about (1) what questions you are most curious about and (2) what questions you can actually test for.

For example, you may have read about the many health benefits of garlic in Mediterranean diets and you wonder, Would a daily addition of garlic to the diet increase the average life span? This is an interesting question, but not one that you can answer with an experiment. However, you could pursue the topic of garlic and health and come up with more testable questions such as these: Does garlic have antibiotic properties? Does garlic expel worms? Does consuming garlic lower blood pressure? These are questions that someone could develop a test for in a controlled experiment. But remember: In order to have several choices of questions you would like to find answers for, you have to record *all* of your questions in your notebook as they occur to you, and decide later which one will form the basis for your experiment.

Some of the questions my students have asked about their plants include:

> Does chayote help lower blood pressure?
> Does raspberry leaf tea help ease menstrual cramps?

Is mint a good breath freshener?
Does eucalyptus oil repel insects?
Is rice water a good fertilizer?

The Hypothesis

Now that you have your question, you are ready to develop your hypothesis. A hypothesis is an educated guess. You make a prediction as to what the outcome of your research will be. It is the assumed answer to the question you asked. This guess is not random — it is based on all the background research you conducted, and it makes sense. The purpose of the experiment is to test the hypothesis and see if it is valid. As with all the other information and steps involved in designing and executing your experiment, remember to write your hypothesis in your notebook.

The following are the hypotheses my students developed for their experiment. They are in the form of answers to the questions in the prior section.

Chayote helps lower blood pressure.
Raspberry leaf tea helps ease menstrual cramps.
Spearmint gum can remove bad breath.
Eucalyptus oil repels insects.
Rice water is an effective fertilizer.

Your experiment may demonstrate that your hypothesis was incorrect — that's fine. The purpose of your research is to test the accuracy of your hypothesis, and if your hypothesis turns out to be incorrect, then you still learned something about your study plant. You may also end up with inconclusive results — that is, results that really can't prove that your hypothesis is true or false. Inconclusive results teach you that additional research, or a change in research methods, is needed. In all cases you are learning from your experiments.

Pennywise Science

A limited budget shouldn't prevent you from conducting research. You probably won't have to look far to see people every day throwing away items that could be salvaged for use in your research. With a little imagination, you can turn ordinary objects into useful items for your experiments. For example, when Ralston needed an insect chamber to test the effect of eucalyptus oil on insects, he figured out a way to construct one out of empty film canisters connected to thoroughly clean, used soda bottles.

Need a container for your plant extract? Thoroughly clean, used baby food jars make good containers for storing extracts.

Are you going to grow seedlings in your experiment? Plastic, styrofoam, and paper cups or containers can be turned into good seedling pots by poking a hole for drainage in the bottom.

And there's no need to buy fabric if you're going to test dyes — if you have some old white t-shirts, you can make cloth swatches out of them for that very purpose.

The Experimental Design

Your experimental design will describe how to conduct your experiment like a step-by-step recipe. Work closely with your teacher to make sure that your experiment tests your hypothesis and is safe. Your methods should be written so clearly that someone else could follow your directions and replicate the same experiment even without your help. Include safety precautions as part of the methodology and take time to write them out. If people are to eat, taste, drink, or apply anything as part of the experiment, explain how this will be done in a safe manner. Before writing your own safety procedures, review the recommended safety precautions at the beginning of this chapter to make sure your experiment meets those standards.

Prepare a list of materials needed for the experiment. Work with your teacher to gather everything you need. You will also need some of the active part of your plant for the experiment. Describe what form of your plant you will use — fresh leaves, dried roots, extract, tincture, and so forth. If you use a Bunsen burner to heat a substance, make sure there are fire extinguishers nearby. Your experiment may require you to wear an apron, safety goggles, and rubber gloves. Make sure you include all safety-related items in your list of materials.

When you write out your experimental design in your notebook, you will include the materials needed, any safety precautions, and the step-by-step procedure.

Your experiment must be designed to test the effect of a change in a condition or environment. To look at this change you need to include the following variables in your experiment.

1. *Subject*, or *experimental subject*. The organism or object you will observe during your study.
2. *Treatment*. This is sometimes called the *manipulated variable*. This is the part of the experiment that you change in each trial. Everything else stays the same.
3. *Control*. The control or *control group* in the experiment gives you a basis for comparing the effects of your treatment on your subjects.
4. *Response*. This is a change or effect that you are measuring in your subject as a result of your test.

Let's clarify the concepts with an example. Luna was studying papaya (*Carica papaya*). After reading that papaya contains enzymes that can tenderize meat and soften the skin, she hypothesized that papaya could tenderize meat. She developed an experiment to test the effect of papaya on beef:

Methods

- Get two pieces of fresh meat that are practically identical (both being the same type and having the same thickness, size, and expiration date) and place each piece on its own plate.
- Rub fresh papaya onto one of the pieces of meat and leave the other piece untouched. Label the plate with treated meat "Treated with papaya" and the plate with untreated meat "Untreated."
- Cover each plate with plastic wrap and refrigerate both plates overnight.
- The next day, use a knife to cut completely through each piece of meat five times (using only the weight of the arm for pressure). Count the number of strokes it takes to cut totally through the meat each of the ten times.

If it required fewer strokes on average to cut through the meat that had been treated with papaya, then Luna would assume that her hypothesis was correct.

Now let's describe this experiment using the experimental terms mentioned above.

1. Subjects: The two pieces of meat.
2. Treatment: The papaya rubbed onto the meat.
3. Control: The meat that had nothing rubbed on it.
4. Response: The difference between how tender or tough the meat was after each treatment, measured as the number of strokes needed to cut through the meat.

Note that in the above experiment, each type of meat was cut through five times. Furthermore, Luna performed her entire experiment several times. These replications help reduce experimental error and make the results more trustworthy. Imagine if Luna had cut through each piece of meat only once: What if she had hit upon a particularly thick or bony section of meat treated with papaya? It would have falsely appeared that the papaya actually made the meat tougher. Since both human error and natural variation exist in experimentation, it is important to repeat the experiment a number of times. Including replications in your experimental design will help to account for those factors and thus improve the reliability of your results. (Luna's poster is shown at left.)

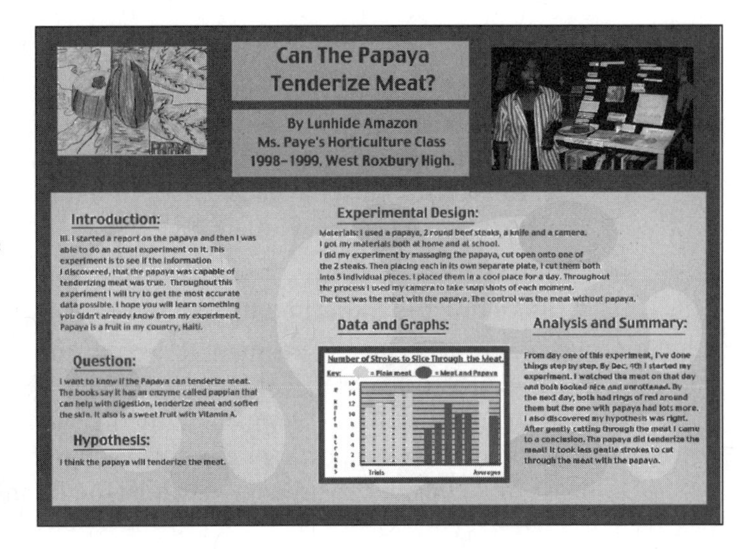

Here is another example of an experiment in ethnobotany, this one conducted by Mia on her chosen plant, chayote (*Sechium edule*), a pale green, pear-shaped fruit in the gourd family that tastes somewhat like cucumber. In Mia's interview with her informant, who was originally from Jamaica, she learned that people in the West Indies eat chayote as a vegetable, and some use it to lower blood pressure. Mia wanted to know whether this was true, and on the basis of the interview and her background research, she hypothesized that chayote can lower blood pressure. She designed an experiment to test her hypothesis.

Methods

- Get chayote and cucumber, wash both, and grate each into its own clean bowl.
- Use a blood pressure monitor to measure the blood pressure of each volunteer subject before the start of the experiment. This is called a *baseline reading*. Record each subject's baseline reading in the notebook.
- Assign half of the subjects to the group that eats chayote and the other half to the group that eats cucumber. Record the groups in the notebook without disclosing their group assignments to the subjects.
- Feed chayote to the subjects in the chayote group and cucumber to the subjects in the cucumber group, without telling them what plant they are eating or what the expected outcome will be.
- Thirty minutes after feeding the plants to the subjects, measure the blood pressure of each subject to see if there is a change from the baseline.

If there was a significant decrease in blood pressure among those who ate chayote compared to those who ate cucumber, then Mia would have confirmed her hypothesis.

What Is a Placebo?

In this experiment on the ability of chayote to lower blood pressure, Mia used cucumber as her control, something to compare with chayote in testing people's response. In this experiment, cucumber also served as a placebo. Often in clinical medical trials, patients are given a placebo which is a pill or medication that resembles the real medicine under study, but has no medical effects at all. The patients do not know if they are receiving the real medicine or the placebo. Sometimes the people taking the placebo show a response or improvement in their health just because they think they are being medicated. By comparing the improvement made using a placebo against the improvement made using the real drug, scientists can determine how effective the medication really is.

Once again, we will describe this experiment using the experimental terms.

1. Subjects: People who ate chayote or cucumber.
2. Treatment group: The people who ate chayote.
3. Control group: The people who ate cucumber.
4. Response: The change in blood pressure among those who ate chayote compared to that among those who ate cucumber.

Conducting the Experiment

If you wrote a clear experimental design, conducting your experiment should be like following a recipe in a cookbook. In your notebook, record the numerical data you collect and the details of what happened during each stage of your experiment. Add the date, time, and other supporting details to each data entry selection. Reflect on problems as they arise and consider them opportunities to add more information to your overall project.

Data Analysis and Graphing

Once the experiment has been completed and all information recorded in your notebook, the next step is to try and figure out what the data means. Scientists use statistics to help them organize and analyze their data. Graphs are used to help organize data into a visual format that is easier to comprehend than a series of raw numbers. Below you will find some descriptions of basic summary statistics and graphs that you may find useful in your data analysis. However, to use these tools you need numerical data.

Whenever possible, you should find a way to *measure* the response in your experiment. Use scientific measuring tools such as a thermometer, clock, timer, scales, ruler, blood pressure monitor, light meter, moisture meter, glucose testing kit, soil testing kit, litmus paper, or any other measuring device to get precise recordings in your experiment. The more exact your measuring device, the more dependable your test results will be.

In the case of Mia's experiment (above), she was able to easily measure the response *quantitatively* (that is, with an objective numerical measure) by using a blood pressure monitor, which gives an objective numerical reading.

But in the case of Luna's experiment, what could she use as a scientific measure of "tenderness"? What exactly is "tenderness"? There is no quantifiable definition, really, so Luna had to devise one so that she could later analyze her data. She came up with a method that measured how many cuts with a knife were needed, with the same pressure applied to each stroke, to pass through each piece of meat.

As we saw in some of our earlier experiments, not all of the data you collect in an experiment are in the form of numbered measurements. For example, if we gather information on people's reaction to a fragrance (as we did in a Laboratory Activity in Unit 6), we might have our subjects check one of five reactions: "Love it," "Like it," "Neutral," "Dislike it," "Hate it." To make it easier to graph and interpret this information we can assign a number to each reaction, from 1 ("Hate it!") to 5 ("Love it!"). The five senses can be powerful data tools if you precisely record your subjects' sensory impressions (and/or your own observations, if appropriate) and systematically assign numerical values to them.

Once you have numerical data (your *data set*), you are ready to analyze it.

Summarize Your Data

Statistical methods can be very complex, but the five types of summary statistics presented here are basic methods that you may already be familiar with: mean, median, mode, range, and frequency. These statistical methods are used to describe a data set so that you can then interpret the results. For example, what was the average reaction that people had to a certain plant fragrance? Or, of two groups of ten seedlings grown under two different conditions, what was the average amount of growth in height in each group in the same time period? Because a few extreme answers affect the overall average, it is also useful to know what the most frequent answers were in an experiment, the range of answers, and the median answer. The five methods are defined in the following ways.

MEAN: The arithmetic average of the numbers in your data set. To obtain the mean, add up all the items in your data set and divide the sum by the number of items in that set.

MEDIAN: The middle item in a data set whose items have been ordered from highest to lowest. To obtain the median, arrange the items in your data set in ascending order of numerical value, then find the middle item (the item with an equal number of items above and below it). If your data set contains an even number of items, take the average of the middle two items.

MODE: The item that occurs most frequently in the data set. To obtain the mode, determine which number in your data set appears more often than the other numbers. Note that a data set can have more than one mode.

RANGE: The difference between the highest number and the lowest number in your data set. To obtain the range, subtract the lowest number from the highest number.

FREQUENCY, or FREQUENCY DISTRIBUTION TABLE: How often a particular value occurs in a given set of data. In addition to pointing out a general trend in the data, it also points out data that does not "fit in." Creating a frequency distribution table is also a logical first step to creating a visual representation of your data.

Let us apply these types of statistical analysis to a student's science project so we can more easily see how it is done.

Ralston wanted to see if eucalyptus (*Eucalyptus globulus*) is a good insecticide and insect repellant. He put ladybugs, grasshoppers, mealworm larvae, and adult mealworms into clean plastic soda bottles with a few air holes. Some of the bottles had a cotton ball soaked in eucalyptus oil. Ralston's control groups were insects placed in bottles with a cotton ball soaked in distilled water. All the insects were left in the bottles for 24 hours. (Ralston and his poster are shown at left.)

For the sake of simplicity we will look only at what happened to the ladybugs in Ralston's experiment. Ralston put 10 ladybugs in the treatment bottle and 10 into the control bottle. At the end of the 24 hours during which Ralston intermittently checked the bottles, all of the ladybugs in the control bottle were alive. Those in the treatment bottle lived 1, 1, 1, 1, 0.75, 0.5, 0.5, 0.5, and 0.5 hours; one of them escaped and lived for more than 24 hours.

The mean or average life span for the control group was 24 hours. The mean or average life span for the treatment group was found by adding all of the numbers and dividing that sum by the number of individuals in the treatment group: $1 + 1 + 1 + 1 + 0.75 + 0.5 + 0.5 + 0.5 + 0.5 + 24 = 30.75 \div 10 = 3.075$. On average, ladybugs exposed to eucalyptus died after 3.075 hours.

The median or middle number in the data set for the control group is 24 hours. The median time the individuals in the treatment group lived was between 0.5 and 0.75 hour, which then averages out to 0.625 hour.

The mode, or most frequently occurring piece of data, for the control group was 24 hours of living; the treatment group was bimodal, with 1 hour and 0.5 hour being the most frequently occurring life spans.

The range, or difference between the highest and lowest numbers, for the control group was 24 - 24 = 0 hours; for the treatment group it was 24 - 0.5 = 23.5 hours.

The frequency for the control group of ladybugs was that the 24-hour life span occurred 10 times. Among the treatment group, a 0.5-hour life span occurred 4 times, a 0.75-hour life span occurred 1 time, a 1-hour life span occurred 4 times, and a 24-hour life span occurred 1 time. We can construct the following frequency distribution table:

Score (hours)	Frequency
0.5	4
0.75	1
1.0	4
24.0	1

How can this data help us to better understand the results of Ralston's experiment? By analyzing what happened, we can *interpret* the results — that is, determine the meaning of them — and make practical suggestions based on the research.

The mean is one of the most critical pieces of data. The huge difference in life span between the control group (whose individuals lived for an average of 24 hours) and the treatment group (whose individuals lived for an average of 3.075 hours) shows that the eucalyptus oil most certainly had a significant negative effect on the ladybugs.

The median shows us that the middle number in the treatment group's data set is much lower than the mean, thus indicating that some of the data may be off balance with the rest of it. (In this case, the escaped ladybug had the effect of distorting the mean.) In the control group the median is the same as the average, which indicates consistency of data for that group.

The mode, or most common number, is also much lower than the mean in the treatment group, whereas it is consistent with the mean for the control group. The range in the control group is 0, whereas in the experimental group there is a range of 23.5. This large range indicates that there was some sort of error or inconsistency with the data (again, due to the escaped ladybug).

The frequency distribution table shows us that 0.5 and 1 hour of living were four times as common as the other life spans. More important, though, the table organizes the data so that it will be easier to graph.

An overview of the data shows us that when including the escaped ladybug in the data set of the experimental group of insects, the reliability of the data is upset. But if we were to remove the escaped ladybug from the data, the results would become much more consistent; since the escaped ladybug did not really "participate" in the full study, it would be legitimate to remove it from the data set. If we do that, the mean for the treatment group goes down to a 0.675-hour life span, the median is a 0.75-hour life span, and the mode is a 0.675-hour life span. The data becomes much more consistent and reliable.

Note that we removed only one piece of data, because that was all that legitimately could be removed: Even though there was only 1 ladybug that lived for 0.75 hours, that ladybug did "participate" in the full study (rather than escaping). If there had been a number of inconsistencies in the data — for instance, if 7 ladybugs died within the first 1 hour and 3 lived for 24 hours — we would have to wonder if something about the design of the experiment or the equipment caused it. Maybe the seal on the bottle broke after one hour and leaked out all the eucalyptus, or maybe the 3 longest-living ladybugs were similar-looking insects that weren't really ladybugs. We would be better off repeating the experiment to try to correct any irregularities that might have been accidentally introduced into the experiment.

Ralston's data show us that eucalyptus may indeed be a powerful insecticide. However, since the ladybug is a beneficial insect that eats aphids and other plant pests, his research would suggest the need for further studies to assess the effect of eucalyptus on a variety of insects under different circumstances before promoting it as an insecticide or insect repellant.

Graph Your Data

An excellent way to look for meaning in your data is to graph it. The visual presentation of your information in the form of pie charts, bar graphs, or line graphs can help you identify the patterns in your data and understand differences between experimental treatments. A brief description and example of each of these three types of graphs follows.

LINE GRAPHS. Line graphs are useful for plotting data that describes changes that take place over time, such as seedling growth per week. Line graphs can also describe a relationship between two continuous variables. For example, you could graph the number of leaves per seedling against seedling height.

The example below shows how much time it took for milk inoculated with *L. acidophilus* to turn into yogurt when mixed with each of several different plant extracts or with no plant extract. The subject is the milk inoculated with *L. acidophilus*; the treatments are the additions of various plant extracts; the control is the inoculated milk with no added plant extract; the measured response is the time it takes the inoculated milk to turn into yogurt (on a scale from 1 to 4, with 1 being watery and 4 being solid yogurt).

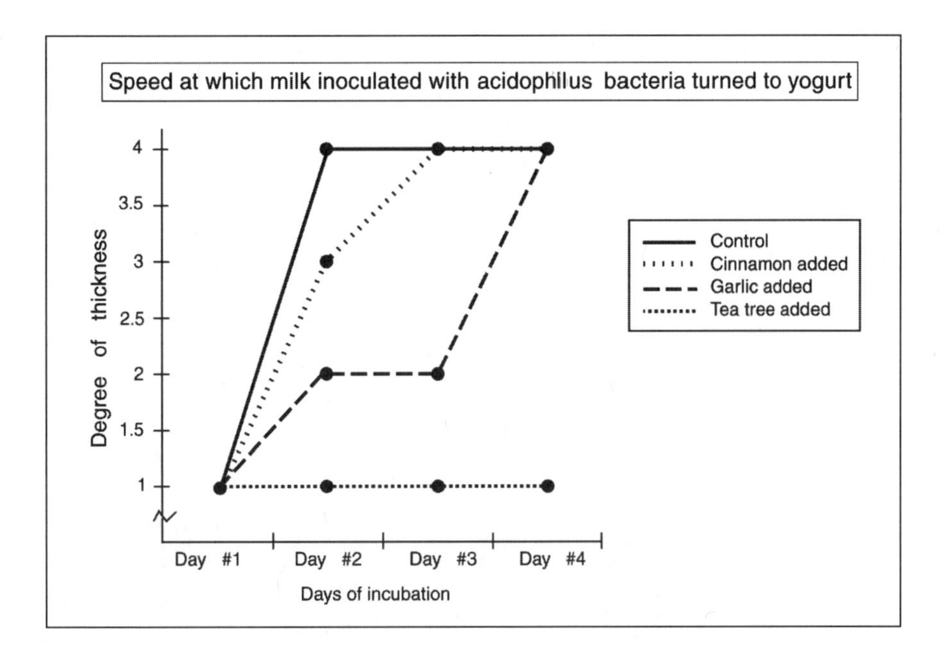

PIE CHARTS. A pie chart is an easy and clear way to show percentages. The entire circle or pie represents 100% of the measured variable. A half circle represents 50%; a quarter circle represent 25%. You can use a protractor to make pie charts by converting the percentages to degrees and then plotting them using the protractor (360° = 100%). To show 45% on a pie chart you multiply 0.45 x 360 = 162°.

The pie chart at the top of the next page shows the data from Michael's study of student preferences for the scents of three different spices. The topic was spice preference. Thirty blindfolded participants were given three different spices to smell without being told what the spices were, and were then asked which of the three they preferred. The presentation of data in the form of a pie chart shows opinion. Note that it does not tell how many people were interviewed, but only the percent of subjects that preferred each scent.

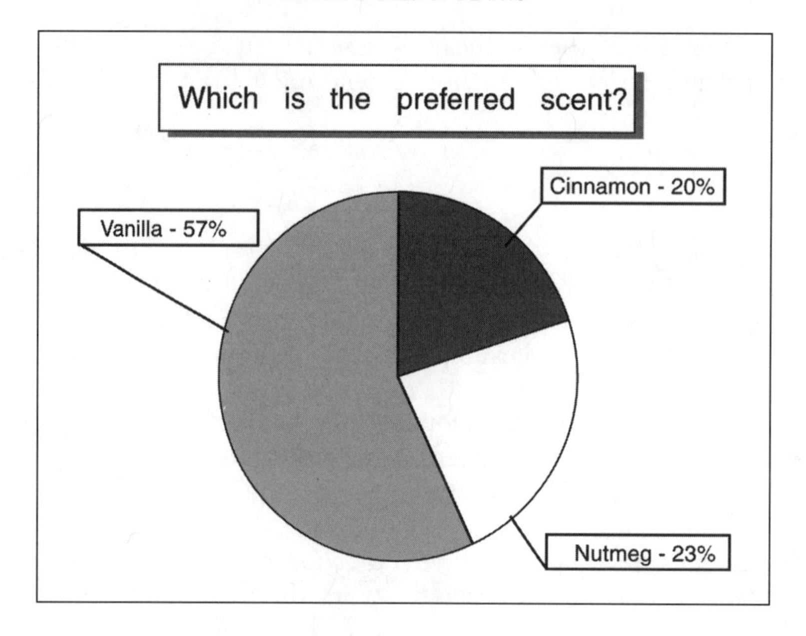

BAR GRAPHS. Bar graphs are useful for comparing different quantities including counts of a given item, or people's responses in a given survey.

Look at the following bar graph to see how strong the fibers were in different types of paper in Nevada's experiment. Nevada made paper out of different plants and measured them to see how strong each one was. She used a spring scale to see how much force was required to rip each type of paper. Her bar graph shows how much force was needed to rip each type of paper.

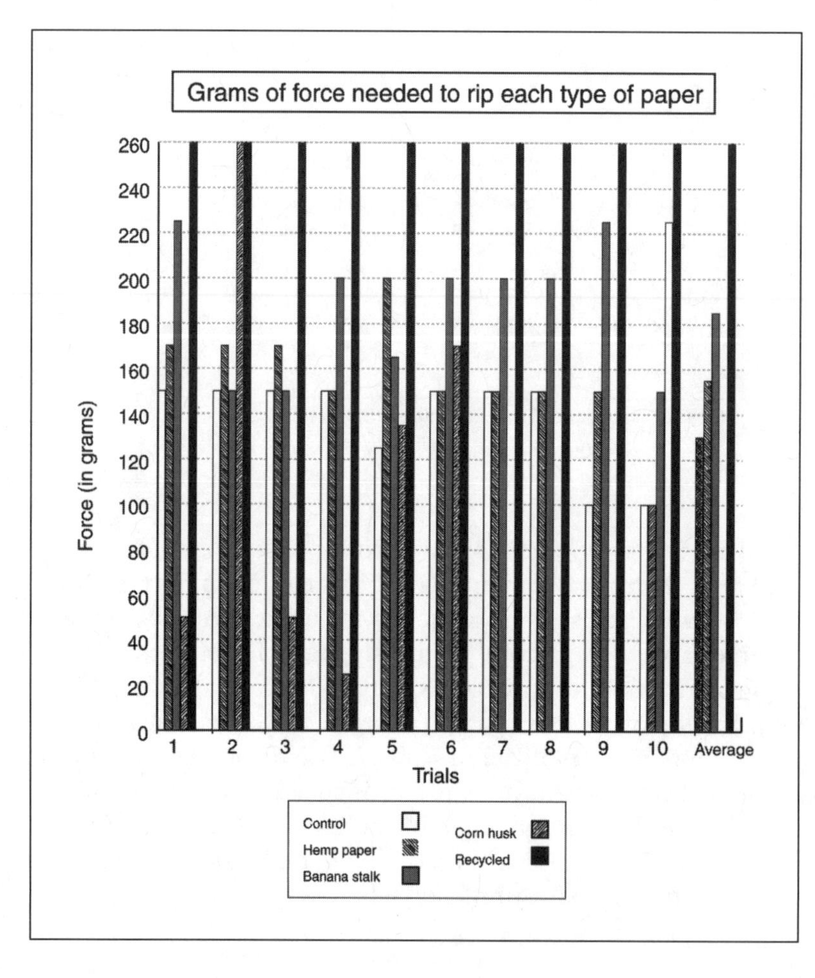

To make your own graphs you will need graph paper, a ruler and a protractor. You may also create your graphs on the computer.

Summary statistics and graphs may be sufficient for you to identify a significant difference between treatment(s) and control in your experiment. On the other hand, it may be that summary statistics and graphs do not allow you to see any clear or statistically significant results from your experiment. You may find that there is great variation in the data you gathered and that the outcome of your experiment is not obvious. In that case, you will need to do further analysis to determine whether your findings are significant. Two additional types of statistical analysis, *standard deviation* and *t-test*, may be helpful in determining whether two data sets are significantly different (in the strict statistical sense) from one another. Your science teacher or math teacher will be able to help you conduct these analyses, or you can consult a web site that describes and explains statistical methods (see Appendix A).

In summary, the following steps may help you interpret your findings and present them in the most simple and clear way possible.

1. Convert non-numerical data to a number set.
2. Summarize your findings using basic summary statistics.
3. Graph your data and look for patterns.

Results and Conclusion

By looking carefully at your graphs and your statistical analysis, you should be able to figure out the meaning of your experimental results. Do the results make sense? Are they what you expected? Did you make an error in experimental design? Did you prove or disprove your hypothesis? Are the results inconclusive? Do you need further research, or a change in procedures? Write everything down in your notebook.

The Abstract

An abstract is a summary of your research. Scientific papers in professional journals almost always begin with an abstract. In one paragraph, or less than a page, the abstract briefly explains what you were trying to figure out, how you tested it, and what your results were. Think of it as a sketch, in that gives your audience a picture of what you are about to explain in greater detail in the body of your presentation.

Presentation

Scientific experiments cannot be considered complete until you organize and share your results. Professional scientists present their findings by writing papers in scientific journals, writing books, or making presentations to their colleagues. In this way, the knowledge and the potential applications and benefits of the research can be used for the greater good. You also will share your results, either by preparing a talk and a poster about your study or by putting your findings on the Internet.

An attractive poster helps to clarify your work and makes your knowledge available to those who see it and hear your accompanying talk. To make a poster you will need posterboard, glue, scissors, and marker pens. Your poster should include the following headings:

- Name
- Class
- Date
- Title of project
- Background information

- Abstract
- Question
- Hypothesis
- Experimental design
    - Materials
    - Safety precautions
    - Subject
    - Treatment
    - Control
    - Measured response
- Data/graphs
- Analysis
- Summary

Under each heading, type or neatly write a summary of what you did. You can get all of this information from your notebook and your background research paper. Photos, drawings, or herbarium specimens of the plant should be added to make the poster more attractive and interesting. Laminating the poster will help it last longer.

For the presentation, bring in some of the materials you used to conduct the experiment, and the plant product, to display alongside your poster. You may even want to repeat some of the experiment during your presentation, to illustrate your procedures. The notebook and the research paper should also be available to your audience, to supplement the poster.

Some people are shy or nervous about giving a talk. In this case it is helpful to practice in front of someone with whom you feel comfortable, and ask them for feedback on your performance. If you are planning to enter your presentation in a science fair or other competition, be prepared to answer some tough questions by judges who want to determine how thoroughly you know your subject and and understand your research. You will need to review everything you learned during the course of your background research and experiment so you will be able to answer those questions and give a well-informed talk. Sometimes the quality of the presentation and the ability of the presenter to answer questions will be the deciding factor in the outcome of a competition.

If you would like to share the results of your research on the Internet, you can e-mail the author for instructions on how to do so: gdpaye@hotmail.com.

Unit 7 Questions for Thought

On a separate piece of paper, answer the following questions as thoroughly as possible.

1. Compare the regulations for approving and developing new pharmaceutical drugs with those for dietary supplements.

2. What are some potential hazards of working with plant materials, and how can you protect yourself from these hazards?

3. Review each of the guidelines for a safe ethnobotanical science experiment. Explain which of them you agree with and which you disagree with and why. (*Note:* You still have to follow them even if you don't agree with them.)

4. What is the purpose of the experimental notebook, and why is it a good idea to add extra things to the notebook such as unrelated observations, ideas, and drawings?

5. Why is background research so essential to designing a good experiment?

6. Why is it still "good science" even if you disprove your hypothesis?

7. Why would an experiment be incomplete if you left out the control?

8. Did Mia or Luna have a better method for measuring the results of her data? Why was one student's method more accurate than the other student's?

9. Determine the mean, median, mode, and range for Luna's data. (Show your math work for each calculation.) The number of strokes needed to cut through the untreated meat (the control group) each time was 11, 12, 12, 14, and 14. For the meat treated with papaya (the treatment group), the number of strokes needed each time was 7, 8, 12, 10, and 10. Show your math work and give your answers in a table like the one below:

	Control group	Treatment group
Mean	____________	____________
Mode	____________	____________
Median	____________	____________
Range	____________	____________

10. Write a frequency distribution table for Luna's data for both the plain meat and the meat with papaya.

11. Create at least two different types of graphs to display Luna's data in two different ways. Make the graphs display different things, such as the median in one and the mean in another.

12. How can you convert opinions from a survey into numerical data?

13. What does it mean to interpret or analyze the data?

14. After completing the experiment, what are the things that you should do to make a powerful, persuasive presentation about what you have learned?

Checklist of Activities for Your Experiment

The following checklist will help you complete all parts of your experiment. As you complete each item, place a checkmark in the space to the left of it. If you see blank spaces as you look down your list, you will need to return to and complete the missing item(s).

Your teacher can also use this checklist as a grading tool, by assigning one of the following numbers in the space to the right of each item: a **4** for "excellent," a **3** for "good," a **2** for "needs improvement," a **1** for "very poor," and a **0** for "not done at all." Your teacher will then add the total score to obtain your grade.

Name: ______________________ Date: ______________________

Topic of experiment: ______________________ Class: ______________________

Did I . . . Score (to be filled in by your teacher)

_____ 1. Get a notebook to record my experimental design and data? _____

_____ 2. Summarize what I learned in my background research in the notebook? _____

_____ 3. Write my question into the notebook? _____

_____ 4. Write my hypothesis in the notebook? _____

_____ 5. Write the experimental design with the subject, treatment, control, and response? _____

_____ 6. Incllude a list of materials needed for the project in my experimental design? _____

_____ 7. Write out the safety precautions for conducting my experiment and check my procedures with the teacher? _____

_____ 8. Determine a way to measure the results of my experiment? _____

_____ 9. Collect all the things necessary to do the experiment? _____

_____ 10. Set up the experiment? _____

_____ 11. Conduct the experiment following the experimental design carefully? _____

_____ 12. Collect accurate and thorough data in the notebook during the project? _____

_____ 13. Record everything that I did and everything that happened? _____

_____ 14. Make more than one trial (the more the better) of the experiment? _____

_____ 15. Determine the mean, median, and mode of my data? _____

_____ 16. Determine the range and frequency of the experimental data? _____

_____ 17. Make at least one appropriate graph that clearly represents the data so that it is easy to understand? _____

_____ 18. Label the graph and make it attractive? _____

_____ 19. Determine the results of the experiment? _____

_____ 20. Write an abstract about the experiment? _____

_____ 21. Make an attractive poster that clearly summarizes the results? _____

_____ 22. Bring some related visual aids or props or the actual experiment to display along with the poster or to attach to the poster? _____

_____ 23. Display the research paper and experimental notebook alongside the poster? _____

_____ 24. Speak clearly about the project and answer questions about it? _____

_____ 25. Demonstrate a solid understanding about the topic of the project? _____

Total score _____

Unit 8

Ecological and Economic Concerns in Ethnobotany

The Importance of Plant Diversity

We have relied on the amazing diversity of plant life as our main source of food, medicines, and material goods since the beginning of human development. Now, as a population of more than six billion, our patterns of resource use threaten the diversity that we depend on for survival. Humans consume nearly 40% of the earth's annual biological production. This intense domination of our ecosystems is eroding the ecological foundations of the plant biodiversity we rely on for so many of our needs. Wild plant species are disappearing at unprecedented rates, and the genetic bases of plants that we cultivate are growing ever more narrow and uniform.

According to a survey conducted by the World Conservation Union – IUCN, one out of every eight plant species in the world is under threat of extinction. This survey, compiled by the World Conservation Monitoring Centre and published as *The IUCN Red List of Threatened Plants*, includes species endangered or vulnerable to extinction and those that are naturally rare. Plant species in limited numbers are inherently vulnerable to any ecological changes, whether it be habitat loss, climatic change, or invasion of exotic species.

In an article in *Harvard Magazine*, the eminent biologist E. O. Wilson warned of the devastation that would be the result of widespread destruction of biodiversity: "The worst thing that can happen — will happen — is not energy depletion, economic collapse, limited nuclear war, or conquest by a totalitarian government. As terrible as these catastrophes would be for us, they can be repaired within a few generations. The one process ongoing . . . that will take millions of years to correct is the loss of genetic and species diversity by the destruction of natural habitats. This is the folly our descendants are least likely to forgive us."

Every new generation starts out with a fresh opportunity to halt habitat destruction, to preserve biological diversity, and to use resources wisely.

Endangered, Threatened, Rare — What's the Difference?

An *endangered* species is one that is clearly at risk of becoming extinct if not protected in the future. A *threatened* species is one that will likely become endangered if not given adequate protection. The terms "endangered" and "threatened" are legally defined terms often used in reference to the Endangered Species Act, enacted in 1973. On the other hand, "rare" is a term that is applied to all plants of conservation concern.

Naturally rare plant species are found in limited numbers due to unique geographic or climatic conditions. Many rare plants can persist in low numbers for many years if not disturbed. However, the fact that their populations are limited makes them extremely vulnerable to ecological disruption.

The countries listed by the IUCN with the most plant species at risk are the United States, Australia, and South Africa. However, the high rates of threatened plant species in these countries are due in part to how much better-known the flora is than in many tropical countries. Although we know that tropical rain forests, as ecosystems, are under threat, we have little detailed information about the particular species at risk or those that have been lost due to logging, mining, plantations, clearing for pastures, cattle grazing, and fire.

Causes of Endangerment

The reasons behind the current rates of plant species loss are many. However, the most significant impact has been that of land-use change. Land-use change, an alteration of the way people use the land, often results in a change in cover — for instance, a forest may be cleared to create a pasture for grazing livestock, or a prairie may be turned into suburban housing developments. Such human activity has substantially transformed one-third to one-half of the earth's land surface. This change has had a profound effect on biodiversity.

The Many Lives of a Single Plant

The saguaro cactus (*Cereus giganteus*) grows in the deserts of Arizona, California, and Mexico. This giant cactus can grow to a height of more than 15 meters (50 feet) and live as long as 150 years. It takes 50 years for this cactus to bear flowers and fruits. Throughout its lifespan the cactus provides shelter and food for numerous organisms.

Carpenter birds and elf owls make nests in the fleshy body of the cactus, and Harris's hawks build nests in the branches. When the cactus blooms, in May, the nectar is enjoyed by bats, doves, butterflies, and bees. In June the fruits are harvested by O'odham Indians, who cook the pulp to make jam, candy, syrup, and wine; curved-bill thrashers, horned lizards, coyotes, and javelina pigs also eat the fruit. Harvester ants gather the seeds. As the cactus nears the end of its lifespan, aquatic beetles swim through the decomposing plant flesh, hister beetles look for fly larvae, and the giant millipede feeds on the decaying plant tissue. When the cactus is dead, it is a home to termites, spiders, giant centipedes, banded geckoes, cactus mice, and spotted night snakes.

When a site is developed for land use, adjacent sites may be affected by *habitat fragmentation*, the reduction of a given habitat type in a landscape and/or the division of remaining habitat into smaller, more isolated parcels. In addition, when land is developed, people, their domesticated plants, and associated animals "squeeze out" the original inhabitants, and often the affected ecosystems are altered by the elimination of one or more *keystone species*. A keystone species is one that plays an important and unique role in the life of an ecosystem; when a system loses a keystone species, its stability falters and other parts of that ecosystem become threatened as well.

A second important reason for the loss of plant diversity is the invasion of exotic species. When exotic or non-native species are introduced into an area, either deliberately or by accident, they may find themselves greatly unhindered by pests or other biological factors that previously held them in check. When such a species begins to grow rampantly and chokes out other species, it is called an *invasive* species.

purple loosestrife (*Lythrum salicaria*)

There are plenty of examples of exotic species that change the composition of our native habitats. For instance, the *Melaleuca* tree, native to Australia, is now displacing native species in the Florida Everglades. Real-estate developers initially planted *Melaleuca* to "dry up" these wetland habitats, but the trees soon spread out of control, out-competing the native species for growing sites. Purple loosestrife (*Lythrum salicaria*) from Europe has also found its niche in our wetlands areas, particularly in the northeastern United States, and, in spite of its lovely purple flowers, is also considered a problem because it displaces native species. The kudzu vine (*Pueraria montana* var. *lobata*) was first introduced to the United States from Japan as an inexpensive and abundant cattle feed, but it soon escaped and has now become one of the most problematic weeds in the southern United States.

A third important reason for the loss of plant diversity is overexploitation. Just like animals, certain plant species can be "hunted" to extinction. As mentioned earlier in this unit, many of our native medicinal plants are being overexploited, which may eventually lead to their extinction. Some scientists believe that the Brazil nut tree (*Bertholletia excelsa*) in the Brazilian Amazon is not regenerating because the seeds are so intensively collected that nothing is left to germinate and make the long, slow trek toward tree adulthood.

Too Popular for Their Own Good

The tremendous increase in the popularity of medicinal plants in the past few years is a mixed blessing for the environment. The expansion in markets for medicinal plants has allowed some farmers to diversify their fields and begin growing medicinal plants using methods that are more sustainable than those typically employed to grow other field crops. On the other hand, the increased demand for medicinal plants has encouraged the overharvesting of certain wild plants, some of which were already at risk due to reduced habitats or their natural rarity.

One grassroots organization in the United States, United Plant Savers, is focusing on the conservation of native medicinal plants in the United States and hopes to create a network of privately held "botanical sanctuaries" across the country where these plants can be preserved and propagated. The following list includes just a few of the native medicinal plants that are currently "at risk" in the wild, according to United Plant Savers.

Native plant (scientific name)	Traditional or current use
American ginseng (*Panax quinquefolius*)	Tonic for treating weakness, stress
Black cohosh (*Cimicifuga racemosa*)	Relieves menstrual pain, menopause symptoms, inflammatory arthritis
Echinacea or coneflower (*Echinacea* species)	Most important immune stimulant in Western herbal medicine; used for all types of infections, including viral infections
Eyebright (*Euphrasia* species)	Used for infectious and allergic conditions of the eyes, middle ear, and sinuses
Goldenseal (*Hydrastis canadensis*)	Used for genito-urinary disorders; antiseptic, laxative, astringent, antibacterial, antifungal
Helonias root (*Chamaelirium luteum*)	Used for female reproductive disorders; encourages regular menstrual cycle; treats ovarian cysts and uterine infections
Kava-kava (*Piper methysticum*)	Stimulant, analgesic, used to treat arthritis (Hawaii only)
Wild yam (*Dioscorea* species)	Anti-inflammatory, used to relieve pain; contains steroidal compounds for relief of menopause symptoms

There is a direct relationship between the loss of biodiversity and the loss of cultural diversity. Indigenous people have been the original stewards of our plant resources throughout the world; their cultures are inextricably tied to the environment. When ethnobotanists work on behalf of plant conservation, they are also working to preserve the original cultural connections to these plants.

The Value of Nature

Before discussing the role of ethnobotany in conservation and development, let's take a more practical look at "what nature is worth." The products and services of ecosystems make up the basic life-support systems of our planet, but they are not accounted for, or given a value, by traditional measurements in our current economic system. They do not show up as part of the Gross National Product (GNP), one of the main indicators used by economists to assess the current health of a nation's economy.

Assigning an economic value to a place, product, or service is a difficult process, but businesses and governments often cannot see the value in species or habitat conservation unless it is assigned a dollar value. They may not understand that mudslides, which cause expensive property damage and loss of life, are more likely to follow powerful storms and hurricanes in areas where soil erosion has been worsened by heavy deforestation; or that air quality decreases, and thus the need for air purification is increased, when green spaces and forests are eliminated from the landscape; or that certain pests can actually become even more of a problem when pesticides that were meant to kill them also wind up killing off their natural predators. For this reason, ethnobotanists are beginning to examine the market value of the plants they are studying, and pointing out the many services and products the plant world gives us.

In a study published in the science journal *Nature* (15 May 1997) a researcher from the Maryland Institute of Ecological Economics drew together the findings of more than 100 studies and estimated the dollar value of the essential ecosystem services that we would have to pay for if they weren't already provided "free" by nature. These include water supply, erosion control, waste treatment, climate regulation, pest control, pollination, air purification, and food production. The total global value of these services was estimated at 33 trillion dollars per year — greatly in excess of the 25 trillion dollars that represented the combined GNP of all the countries in the world for that year. Although it may be more immediately profitable to use natural resources in whatever manner we wish, in the long run, greater wealth — in very real economic terms — is created by conservation and wise use of natural resources.

The Role of Ethnobotany in Conservation and Development

Once businesses and governments fully understand in economic and ecological terms what is at stake if we do not conserve natural resources and use them wisely, it becomes the responsibility of ethnobotanists and other scientists to determine the best ways to plan projects in conservation and development.

Using Invasives to Spare the Endangered

In the medicinal herbal community and among such groups as United Plants Savers, there is a movement to switch from using overharvested medicinal herbs to using invasive or overabundant species that have medicinal properties. For instance, eyebright (*Euphrasia* species), an at-risk plant, is used by herbalists for eye and ear infections and sinus allergies. Some research indicates that the overabundant purple loosestrife (*Lythrum salicaria*) may have similar medicinal properties and thus has potential as a substitute for eyebright. Not only would this help the conservation of eyebright — it would also be a productive way to start ridding our wetlands of the pernicious loosestrife so that these ecosystems can return to their normal functioning state.

Since ethnobotany became its own academic discipline, ethnobotanists realized that their research could make a significant contribution to sustainable development and the conservation of natural resources. Conservation is particularly difficult in the tropics, where so much of the plant life has yet to be catalogued and where a scarcity of funding and other resources impedes the development and maintenance of national parks and nature reserves. So then, if plant conservation is to take place (particularly in the tropics but also elsewhere), it will have to begin on the multitude of individually owned plots that make up the majority of the earth's land holdings. When ethnobotanists study the skills that many small farmers in the tropics have been practicing for generations, and study the ways that other indigenous peoples use and substainably manage their plant resources, we gain a great deal of insight into the simultaneous conservation and sustainable use of plant resources that have been practiced for generations.

Small-scale traditional agriculture in the tropics is often characterized by systems of mixed planting (a practice known as *polyculture*) and the combination of wild, cultivated, and semi-cultivated plants in the same plot. These farmers maintain species-rich, genetically diverse fields that rely minimally or not at all on the use of agricultural chemicals. These farmers are often familiar with local plants they use as food and medicine or in pest control.

Studying how local people view, use, and manage resources brings a rich body of knowledge to the processes of project planning and development. This particular area of study is often called *ethnoecology*, and it is increasingly recognized as an important part of successful conservation and development projects. Indigenous people are more likely to support and participate in a conservation plan that incorporates their knowledge and skills and recognizes their right to participate in planning, decision-making, and implementation.

As the twentieth century drew to an end, there began a movement not only to respect indigenous knowledge and incorporate it wherever possible, but also to compensate and credit indigenous peoples if their knowledge leads to financial gain. Traditional healing has long been an area of interest to *pharmaceutical prospecting* — that is, searching for bioactive compounds for use in commercial drugs — but it has not always been the case that healers were compensated for their part in educating researchers about the useful plants that then became patented, profitable, commercial drugs. Now, however, there is greater emphasis on *intellectual property rights* — the notion that a person who develops a body of unique knowledge or an idea should have some control over how that intellectual property is used and should receive a fair share of the profits that come from it.

This movement toward respecting the diversity of cultures and the rich traditional knowledge they possess — along with a more just distribution of financial gains that result from that knowledge — was first expressed in the Declaration of Belém. Created in 1988 by the First International Congress of Ethnobiology, this declaration emphasized the importance of biodiversity to indigenous cultures around the globe and their right to consult and participate in projects that affect them or their environment. Because this statement was the first one to be made by a major academic organization, it is a significant one. The declaration reads as follows:

The Declaration of Belém

As ethnobiologists, we are alarmed that:
SINCE

- tropical forests and other fragile ecosystems are disappearing,
- many species, both plant and animal, are threatened with extinction, and
- indigenous cultures around the world are being disrupted and destroyed;

and GIVEN

- that economic, agricultural, and health conditions of people are dependent on these resources;
- that native peoples have been stewards of 99 % of the world's genetic resources; and
- that there is an inextricable link between cultural and biological diversity;

THEREFORE, as members of the International Society of Ethnobiology, we strongly urge action as follows:

1. Henceforth, a substantial proportion of development aid be devoted to efforts aimed at ethnobiological inventory, conservation, and management programs.
2. Mechanisms be established by which indigenous specialists are recognized as proper authorities and are consulted in all programs affecting them, their resources, and their environment.
3. All other inalienable human rights be recognized and guaranteed, including cultural and linguistic identity.
4. Procedures be developed to compensate native peoples for the utilization of their knowledge and their biological resources.
5. Educational programs be implemented to alert the global community to the value of ethnobiological knowledge for human well-being.
6. All medical programs include the recognition of and respect for traditional healers and the incorporation of traditional health practices that enhance the health status of these populations.
7. Ethnobiologists make available the results of their research to the native peoples with whom they have worked, especially including dissemination in the native language.
8. Exchange of information be promoted among indigenous and peasant peoples regarding conservation, management, and sustained utilization of resources.

Extinction Is Not a Temporary Condition

One important point must be emphasized over and over. Species loss is the one wholly irreversible component of global environmental change. Global climate change will change the distribution and abundance of organisms on the planet, but the current rates of species loss will determine what is left to "redistribute" and recover.

Human population growth and our patterns of resource use have a big impact on species loss. In 1900 there were about 1.5 billion people on earth; in 2000 the population reached 6 billion — a fourfold increase in the number of people who make demands on natural resources such as food, water, medicine, minerals, oil, and wood. The increased use of energy in industry and agriculture makes a particularly strong impact on the environment. However, we can make a difference by promoting sustainable resource use through the choices we make, whether we're choosing what to eat for dinner, what to grow in the garden, what career to pursue, or how many children to have. As ethnobotanists, we also know that traditional knowledge about plant use and management of natural resources plays a vital role in the overall conservation effort, now and in the future.

Unit 8 Questions for Thought

On a separate piece of paper, answer the following questions as thoroughly as possible.

1. Which environmental problem is the most irreversible and why?

2. Describe how the following human activities threaten plant species:
   a. land development
   b. transfer of exotic species to a new place
   c. overexploitation
   d. human population growth

3. What causes some species to become invasive while others become endangered?

4. What is a possible solution for controlling exotic, invasive species?

5. What can be learned about environmental practices from indigenous people?

For questions, 6 through 9, think of one of the services that nature performs for us — clean air, pollination, water filtration, pest control, and so forth — and answer the questions with your chosen service in mind.

6. What service did you choose, and what would happen to us if nature could no longer perform this service for us?

7. If we had to pay for this service, how much do you think it would cost per year for each person?

8. What human activities may threaten this service that nature performs?

9. What human activities could help to protect this resource and service?

10. How does the Declaration of Belém propose that we protect species and treat native peoples?

Unit 8 Activities

In the lab activities in this chapter you will learn about some of the ecological methods that are commonly used to estimate plant diversity and abundance. You will also learn about estimating market value for the plants in your quadrats. If the plant you have been studying throughout this book grows wild in your area, you may find it in your quadrats. However, if your study plant does not grow in the wild in your area, you can still conduct these experiments and learn more about ecological methods.

Field Exploration: Create a Study Plot and Map It

Name: ______________________________ Date: ____________________

Because entire ecosystems are too large to be studied in truly useful detail, ecologists usually select a representative area for their investigations of living things and the environment. These selected study sites are sample areas called *plots*. Plots may be divided into smaller sampling units, called *quadrats*, that are square or rectangular in shape. For example, a 1-hectare plot might be divided into 25 quadrats of 20 × 20 meters each. Circular sample areas are also employed in some ecological studies.

To sample trees in a forest, ecologists often use a 100 × 100 meter square sample area. Small trees and shrubs would be studied in a 10 × 10 meter square area, and herbaceous plants and grasses studied in a 1 × 1 meter square area. A study area could also contain numerous samples located along a straight line or transect.

In this activity you will select and lay out a quadrat for your study area. This quadrat will be used for the subsequent laboratory activities that follow.

Caution: Avoid any harmful plants (such as poison ivy) or harmful animals in your study area during this activity. Consult a field guide for your region in order to determine what harmful plants or animals you are likely to encounter and to find out how you can identify and avoid them. Bring a first aid kit along with you for emergencies.

Materials needed

Measuring tape (at least 10 meters long)
4 wooden stakes for each study plot
Hammer
40 meters of cord or string for each quadrat
Graph paper and a clipboard to write on
Colored pencils
Field guides to identify species in the plot

Directions

1. Select a study site that is accessible and has a representative sample of local plant life and at least one tree.

2. Use the measuring tape to measure a square plot 10 × 10 m.

3. At each corner of the plot, hammer one of the wooden stakes into the ground so that about 20 cm is left above ground.

4. Tie the string onto one of the stakes and carry it to each of the four corners of the plot, tying it to each stake as you go.

5. If you wish to make more than one quadrat, repeat steps 1 through 4 for each one.

6. Using the graph paper, draw your study site. Determine the scale. For example, 2 cm on the map could equal 1 m on the actual plot. Write down the scale and then draw the outline of the plot on the graph paper.

7. Write down your name and the location of the study site.

8. Devise a key to help you identify the species in your plot. For example, a green maple leaf could represent a sugar maple tree, a red maple leaf could represent a red maple tree, and a vertical purple line could represent purple loosestrife. Put the key outside of the map and make a symbol for each type of organism you find in the study plot. If you cannot identify some organisms, use a field guide.

9. Using the symbol from your key, add them onto your map showing the location and number of individuals of each of the important species that you find in your plot. Also, draw any non-living features in your map, such as a rock, trash can, puddle, fallen tree, and so forth. Avoid counting each individual grass plant (unless this is your special interest) since these are usually too numerous to count. Add the numbers of all the other plant life to your map.

10. Carefully turn over rocks and logs to find insects, reptiles, and amphibians, remembering to replace them as you found them. Look for evidence of these and other animals by searching for tracks, droppings, holes in the ground and trees, hair, nests, sounds, and so forth. Draw all of these onto your map (with a symbol from your key).

Optional activity
If you have access to GIS (geographic information systems) mapping software, try plotting your map on the computer. Mapping software allows you to create maps that indicate elevation, landforms, or any other data you wish to include in your map.

Analysis

1. Which organism is the most plentiful one in your study plot? ____________________

2. Why do you think this organism is so common in this particular ecosystem?

____

____

3. What were some of the other organisms you found in your study plot?

____

____

____

4. Describe at least one food chain that you saw in your study plot.

____

____

____

5. How would you best describe this study plot? Circle all that apply.

urban	suburban	freshwater stream	freshwater lake	park
rural	salt water	freshwater pond	vernal pool	field
desert	tropical climate	undisturbed natural area	disturbed habitat	lawn
estuary	temperate zone	deciduous forest	intertidal zone	beach
landfill	grassland	coniferous forest	tropical forest	

disturbed habitat undergoing succession other: ____________________

6a. Is the plant you have been studying present in this study plot? ________
6b. If so, how many individuals are there? ________
6c. Is your plant abundant, rare, or nonexistent in this habitat? ____________________
6d. Can your plant grow wild or must it be cultivated? ____________________

7a. If you were to suggest harvesting one plant in this study plot for economic uses, which plant would you select? ____________________
7b. Explain why you selected it. ____________________

____

____

7c. In what ways could it be used to help people? ____________________

____

____

8a. In which month and season did you conduct this study? ____________________
8b. How do you think your study plot might change over the course of a single year?

____

____

____

____

8c. How do you think your study plot might change over the course of the next ten years?

____

____

____

____

Field Exploration: Calculate the Abundance of Plant Life in Your Study Plot

Name: ______________________ Date: ______________

To calculate the relative abundance of each species of plant life in each study plot, use the following formula:

1. Count the number of each type of plant.

2. Add all the plants to find the total number of all the plants in the study plot.

3. Determine the abundance of each plant type as a percentage of the total number of plants in the study plot.

For instance, if there are 2 mullein (*Verbascum*) plants and a total of 200 plants (not including grass) in the study plot, the abundance of mullein would be calculated as follows: $2 \div 200 \times 100\% = 1\%$. The abundance of mullein in this study plot is 1%.

Analysis

1. The total number of non-grass plants in this study plot is ____________.

2. List the number of each type of plant in the study plot and then estimate the abundance of each using the formula given above.

Plant name	Number in plot	Formula	Abundance
____________	________	____________	______%
____________	________	____________	______%
____________	________	____________	______%
____________	________	____________	______%
____________	________	____________	______%
____________	________	____________	______%
____________	________	____________	______%
____________	________	____________	______%

3a. Which plant in the study plot is the most abundant? ______________________

3b. Which plant in the study plot is the least abundant? ______________________

4. Based on this activity, which plants do you think should be protected and which could be used to a greater extent? ______________________________________

____

Field Exploration: Write a Journal and Ask Questions about Organisms in Your Study Plot

Name: ________________________________ Date: ____________________

All good science comes out of asking questions. Once an interesting question is asked, the job of science is to come up with a method for answering the question. If you look at the plants in your study plot or walk around your school or community, you are sure to come up with a series of questions. For instance, you might wonder, Why are the plants bigger on the south side of the study plot than on the other sides? How could I repress the growth of nonnative invasive species while promoting the growth of rare or threatened species? How do the Canadian geese around the study plot affect these plants? These questions, and many others that you will think of, arise out of careful observation and thought.

In this activity you will keep a nature journal and write down your observations of what you observe in and around your study plot for at least a week. You will also formulate questions that you have about the things you've seen, and you will determine an approach you could use to find an answer to that question.

Materials

Small notebook
Pen or pencil
Your study plot, or another outside area you could observe

Directions

1. Go to your site and carefully observe it for about 10 minutes a day for at least one week.

2. While watching your site, write down in your notebook or journal what was happening at your site and your observations of it. Do not wait to write your notes — you might forget important details. Look for evidence of human activity or intervention. Look for evidence of animal life such as tracks, droppings, fur, feathers, and so forth. Look at how much sun or shade there is at different times of day. Look at the slope of the land. How wet or dry is the soil? All entries should include the date, location, and time. Entries could also include some or all of the following:

- latitude and longitude of the site
- abundance of different species
- types of vegetation
- health of vegetation in the plot
- presence or development of flowers and fruits
- interactions between organisms
- weather conditions
- signs of pollution
- questions that arise in your mind
- drawings of what you observed

3. After a week or two of writing down your observations and thoughts, write down what the biggest question was to arise from your observations.

4. Working by yourself or with a group, design an experiment that would help answer your question. Write the experiment in your notebook.

5. If you decide that you want to pursue an answer to your question, you can find details on how to set up an original control experiment in Unit 7 of this book.

Field Exploration: Survey Local People about the Marketability of Plants in Your Study Plot

Name: ______________________ Date: ______________

In this activity you will survey three people from your community about the more dominant, common, and useful plants in your study plot (and the area surrounding it) and about the uses of these plants. These people should be people who know something about the local habitat and natural history as well as uses of the plants. This type of survey is frequently conducted by ethnobotanists to determine whether there would be a larger possible market for a plant than is presently being maximized. You should only interview people about plants that are common or abundant in your study plot or the area surrounding it, since you would only be interested in marketing commonly available plants while leaving rare plants undisturbed.

Materials needed

3 copies of the survey form that appears on the following page

Herbarium specimens (see Unit 3 for instructions) of the more abundant and/or useful plants in your study plot and the surrounding area

Directions

1. Find a local person who is knowledgeable about local natural history.

2. Show the person the voucher specimens of the plants in the study plot and ask him or her to complete the survey about the plants that he or she thinks are the most important and useful.

3. Find two other local people and repeat steps 1 and 2 with each of them.

4. Complete the chart and questions below to analyze your results.

Analysis

1. Use the chart below to record the results of survey question 3, where respondents ranked each plant's usefulness on a scale of 1 to 5. Add up the scores to obtain the total. Rank each plant from most to least valuable.

Plant	Respondents' rankings A	B	C	Total score	Rank
____________	___	___	___	_____	_____
____________	___	___	___	_____	_____
____________	___	___	___	_____	_____
____________	___	___	___	_____	_____
____________	___	___	___	_____	_____

2. Which plant did respondents think was the most valuable? ______________

3. Is this a common or abundant plant in your study plot? ______________

4. Based on the results of this survey and the studies of how abundant the plants in your study plot are, which plant(s), if any, do you think might have economic value and might be able to be marketed to people in your community?

____

____

Marketability Survey Form

Name of person interviewed: ____________________ Age: ____ Date: ___/___/___

Plant A: ____________________________

1. Are you familiar with this plant? Yes _____ No _____
2. How is this plant used (circle all that apply)?

   Food Fiber Fuel wood Building material

   Medicine Grazing Other (describe) ____________________
3. On a scale of 1 to 5, with 1 being least valuable and 5 being most valuable, how useful would you say this plant is to humans (circle one)? 1 2 3 4 5
4. Do you think there would be a local market for selling this plant (check one)?

   _____ It could be sold raw directly to the public

   _____ It could be sold dried directly to the public

   _____ It could be sold raw or dried directly to the public

   _____ It could be sold to a manufacturer for processing

   _____ It is not a valuable plant and would not make any profits
5. How common and abundant do you think this plant is in this area (circle one)?

Overabundant Very abundant Common Rare Endangered

Plant B: ____________________________

1. Are you familiar with this plant? Yes _____ No _____
2. How is this plant used (circle all that apply)?

   Food Fiber Fuel wood Building material

   Medicine Grazing Other (describe) ____________________
3. On a scale of 1 to 5, with 1 being least valuable and 5 being most valuable, how useful would you say this plant is to humans (circle one)? 1 2 3 4 5
4. Do you think there would be a local market for selling this plant (check one)?

   _____ It could be sold raw directly to the public

   _____ It could be sold dried directly to the public

   _____ It could be sold raw or dried directly to the public

   _____ It could be sold to a manufacturer for processing

   _____ It is not a valuable plant and would not make any profits
5. How common and abundant do you think this plant is in this area (circle one)?

Overabundant Very abundant Common Rare Endangered

Plant C: ____________________________

1. Are you familiar with this plant? Yes _____ No _____
2. How is this plant used (circle all that apply)?

   Food Fiber Fuel wood Building material

   Medicine Grazing Other (describe) ____________________
3. On a scale of 1 to 5, with 1 being least valuable and 5 being most valuable, how useful would you say this plant is to humans (circle one)? 1 2 3 4 5
4. Do you think there would be a local market for selling this plant (check one)?

   _____ It could be sold raw directly to the public

   _____ It could be sold dried directly to the public

   _____ It could be sold raw or dried directly to the public

   _____ It could be sold to a manufacturer for processing

   _____ It is not a valuable plant and would not make any profits
5. How common and abundant do you think this plant is in this area (circle one)?

Overabundant Very abundant Common Rare Endangered

Field Exploration: Determine the Market Value of the Plants in Your Study Plot

Name: ______________________________ Date: __________________

In this activity you will go to a local supermarket, ethnic market, health food store, or drug store to see if any of the plants found in your study plot are currently being sold, either by themselves or as an ingredient in some product. Look at foods, teas, herbs, herbal preparations, cosmetics, and other products. You will try to determine the price per unit of the plant in the product. This information, along with the information you get from your survey in the previous activity, will help you determine whether there might be a market for selling your plant. If the plant is not currently being sold, there may still be ways to utilize it and profit from it, although it might be more difficult to create a new market than to break into an existing one.

Materials needed

The survey form that appears on the following page
Local supermarket, ethnic market, health food store, or drug store

Directions

1. Look for the plants from your study plot being sold whole or as ingredients in other products in the market.
2. Record your findings on the survey sheet, below.
3. Analyze the results using the following questions.

Analysis

1. Were you able to find evidence of any of the plants in your study plot being sold in the store you visited? ______________________________

2. How were the plants in your study plot being used? ______________________________

____

____

3a. If it was being sold, how much do you estimate the plant was being sold for per unit (state the unit)? ______________________________

3b. How does the price of this plant compare to the price of a similar quantity of bananas or potatoes? ______________________________

3c. How does the price of this plant compare to the price of a similar quantity of herbal tea? ______________________________

4. Based on this activity, do you think one of the plants from your study plot might be profitable if harvested in a sustainable way? ______________________________

Survey Form for Plant Products Sold in a Local Market

Date: ________________ Location: ______________________________

Name of store: ____

Type of store (health food, Chinese, Hispanic, etc.): ______________________

1. Names of plants from my study plot that were being sold at the store whole or as an ingredient in another product: ______________________________

____

2. What are the products that are being sold? ______________________

____

____

3. How is each of these plants sold (fresh, dry, preserved, tincture, etc.)?

____

4. Are there other ingredients in any of the products with the plant? If so, what are they?

____

____

5a. In what units of measurement are the products with plants sold (ounces, pounds, grams, gallons, liters, etc.)? ______________________________

5b. What is the cost of these products per unit? ______________________

6. Where are these products manufactured? ______________________

7. Is there a particular time of year when each product is not available?

____

8. Is there any other relevant information about the store and its products?

____

____

Unit 9

Supplemental Activities

In this final unit, we complete our ethnobotany study by looking at ways to celebrate and share what we've learned. As a way to continue this work, we also explore career and volunteer opportunities related to botany, conservation, and environmental science.

Ways to Celebrate and Share What You've Learned

Host a Multicultural Festival

As a grand finale, your class may want to host an ethnobotanical, multicultural festival at your school or in someone's home. Invite family members, administrators, friends, and your informants to come and see some of what you've learned about plants. As part of the festival you might share ethnic food, conduct a plant-based ritual, play games, make and distribute plant-based gifts, or give project presentations. These activities are described below.

Have a Feast

Parties and festivals always include food. If you plan an ethnic food potluck, each person could share a dish from their own culture. Participants could bring along a recipe to accompany their food and post it next to the dish. You could also create a class cookbook with the recipes and include a sprinkling of the most interesting ethnobotanical pieces of information each person learned about their plant. You might even want to sell copies of your cookbook as a class fundraiser, or use the proceeds to make a donation to a conservation or botanical organization the class members have agreed on.

Plan a way that your class can help the guests learn about the multitude of plants they will be serving up on their plates. Some questions you might post to the participants in the festival include: Which dish contains the most plant species? Which plant or plant family is the most common in the potluck? Which dishes contain products from trees? Which dishes contain products from tropical plants? Which dishes contain products from Central America (or other specific areas of the world)?

Hold a Ritual

Rituals form a part of every culture and many are plant-based rituals. Consider a Hawaiian luau with poi to eat and leis to wear, a Japanese tea ceremony, an Indian wedding with flower garlands, a Jewish Sukkot meal in a sukkah made of corn stalks and gourds. Consider the presentation of flowers at funerals, weddings, and Valentine's day; or the planting of trees in memory of someone or to mark a special occasion. Your festival might include one complex ritual that involves plants, or several smaller demonstrations of rituals from around the world. Make sure the ritual(s) you conduct are legal, safe, not offensive to anybody, and inclusive of all the cultures or religions represented in your class.

Present Your Work

The festival offers a prime opportunity for project presentations. Poster displays could decorate the room and informal talks be given to accompany them. You could also create a play, a song, a rap, a computer slide show, or a video presentation about what you have learned. Don't forget to invite the informants that inspired each person's original choice of plant; informants could even be invited to speak and share some of their knowledge or stories.

Make Plant-based Gifts

Create herbal shampoo, potpourri, flower arrangements (Nevada is shown here with some of her creations), wreaths, corsages, and other small gifts to present to your guests. The preparation of these items might even form part of the festival activities, if someone in the class made a poster with instructions (such as the one at the beginning of this unit, made by Jerry to explain the steps involved in creating a flower arrangement). In the Activities section of this unit you will find directions for making some of these items; there are also many books that provide instructions for plant-based and floral gifts (you can find some of these books listed in Appendix A). Your class can come up with their own plant-based gift ideas, too.

Play a Game

If younger children attend your festival you can modify traditional games to introduce botanical concepts. For example "tag" can become "plant tag." In this game, a group of youngsters (or oldsters) are introduced to five common plants in the yard. To play, the person who is "It" calls out the name of a plant. The plant they name becomes "base." If you run and touch that plant you are safe, but if you're off base you can be tagged "It." As people get to know the plants by name, more species can be added to the game.

Plant a Garden

A fun and educational but long-term project is to create a garden with an ethnobotanical theme. Your garden could be a multicultural one and incorporate plants that celebrate the different ethnic groups represented in your class. (At left, Donna holds one of the plants she propagated to represent her Italian heritage.) Or the garden could highlight the plants of one particular ethnic group such as the Native Americans that originally lived in your area. Alternatively, your garden could focus on creating a mood or an enjoyable space. There are themes your garden could aim for, such as traditional crops with historical significance, medicinal plants, and wild foods. An heirloom garden, for instance, includes plants that have been passed down through generations; heirloom varieties are usually excellent cultivars that have been overlooked simply because some gardeners became attracted instead to hybrid or otherwise manipulated varieties that stressed quantity rather than quality. By growing and saving seeds of heirloom or rare plants, you can have a direct impact on biodiversity preservation. You could also plant a garden with a conservation theme rather than one that emphasizes only ethnobotany. For example, you could design a garden to attract butterflies, birds, or other wildlife.

Consult gardening and landscaping books to help you with the planning and design of your garden. It may be helpful to keep a garden notebook or calendar to remember what to do next in the garden and what has already been done. You can also use this notebook to create a garden-based experiment.

Some things to consider when planning and planting include the following:

- If it is an educational garden, put water-resistant labels on each plant, including the common name, scientific name, and features of interest.
- Select plants that are well-adapted to your climate.
- Arrange for care and maintenance of your garden during vacations and holidays.
- Determine what tools and supplies you will need and where to store them.
- Make plans for soil enrichment and consider starting a compost pile.
- Know the growth requirements for each plant.

Making the World a Better Place: Careers Related to Plants, People, and the Environment

Most people want to make a difference in the world. In writing this book it is my hope that activities and information presented here might inspire you to learn more about the relationships among plants, people, and the health of our planet. Perhaps you will want to study ethnobotany or environmental science in college, or turn your interest in growing plants into a full-time job.

If you're concerned about the condition of our environment, you don't have to wait until you're an adult to start making an impact on the world. You may want to begin working as a volunteer for an environmental group or botanical garden. Or you may start a conservation club at school. In the words of one student: "If I wait until I'm grown up to do something to protect endangered species, there won't be anything left!"

This section of the book briefly mentions some career opportunities related to plants and the environment. We include general fields or career areas that involve the study, production, marketing, and conservation of plants, and careers related to environmental education and communication. Some jobs require vocational training in high school, other careers require a Ph.D. Although we're limited here to a general list, we hope it will spark enough interest for some students to explore these careers through the library, via the Internet, or with a guidance counselor. We also mention a few ideas for volunteer opportunities in related fields.

Careers in Plant Production and Management

The fields of *agriculture*, *horticulture*, *forestry*, and *landscape design* all involve people directly in the production, protection and management of plants for human use and enjoyment, from newly planted pansies to ancient redwood forests. In each of these areas there are many different job opportunities, some that require little or no college education and others that require advanced degrees. If you relish working in an outdoor setting surrounded by plants, you might consider finding out more about one of these fields of work.

Environmental Education

A broad variety of careers involve educating people about our environment. Without an informed public we have little hope of making changes. We need motivated *science teachers* at the elementary level; and *biology, horticulture, and environmental science teachers* in the secondary schools. *Naturalists* teach environmental education to people directly in camps, nature preserves, and state and national parks. *Educators in museums, botanical gardens,* and *zoo*s provide hands-on workshops and presentations to the student groups and visitors that come to these institutions each year for specialized learning. *Museum curators* also create educational exhibits. *Agricultural extension agents* advise farmers and gardeners about the latest developments in plant production.

Making Decomposition Come Alive

Gray Russell is the Compost Project Manager of Bronx Green-Up, a community outreach program of The New York Botanical Garden. Gray worms his way into the hearts and classrooms of New York City schools to make the study of compost come alive — literally. Armed with a large plastic box full of vegetable scraps, strips of wet newspaper, and a few handfuls of red worms, Gray explains how just one pound of these wriggling recycling machines can turn 200 pounds of apple cores, carrot tops, and other vegetable scraps into the perfect fertilizer for houseplants and garden plots.

Since its inception in 1993, the Compost Project has taught composting workshops to teachers in hundreds of New York City schools. These teachers then go back to their schools and teach their students that 20% of solid waste in New York City comes from kitchen scraps and yard waste (such as grass clippings and fallen leaves), that composting those materials means they won't end up in dumps, and that composting isn't hard to do.

A little-known but interesting career is *horticultural therapy*, which uses the activity of cultivating live plants to heal and rehabilitate people with injuries, physical disabilities, or emotional problems.

Environmental Journalism and Visual Arts

Another way to increase public awareness about the environment is through your writing and artistic talents. *Environmental journalists* keep the public informed on a host of issues, from the dumping of toxic waste to deforestation in the Amazon. Effective reporting may be the key to moving the public to action on conserving critical habitat or making people aware of pesticide contamination in local water supplies.

The visual mediums of *photography*, *video production*, and *website design* likewise can be powerful tools to promote environmental awareness, education, and action. The creation of books and educational materials also provides career opportunities to people with artistic talents. *Botanical or biological illustration* is a specialized skill necessary for scientific publications.

Environmental Law

Lawyers and *politicians* concerned with environmental issues can make a major impact by working to change or enforce laws related to environmental protection. The passage of legislation to save wetlands, protect endangered species, and promote recycling, for example, result from the combined efforts of people with skills in *environmental law*, *research*, and *community outreach*.

"Green" Business

People with a background in *business* or *economics* can work on behalf of the environment by helping to market or create markets for products that have been grown or manufactured using sustainable methods. Groups that deserve particular mention are *alternative marketing* organizations (such as Oxfam, Ten Thousand Villages, and Novica) that sell goods produced by indigenous peoples or small cooperatives in developing countries, making sure the producers receive a fair price for their efforts. The sale of handicrafts or other useful items using locally available materials help support local economies.

The "Shady" Side of Coffee

Where do migratory birds go when winter descends upon us? Many of our migratory birds spend their winters in the Central and South America. Given the high rates of deforestation, however, these birds increasingly are taking refuge in the forest-like environment of shade-grown coffee.

Originally from Ethiopia, coffee (*Coffea arabica*) is a small tree adapted to grow in a forest understory. Traditional coffee plantations mimic forest conditions and play a key role in protecting biological diversity in Latin America. The problem is that shade coffee plantations are now a threatened habitat.

Over the past two decades, small farmers have been encouraged to "modernize" — to switch to sun-tolerant coffee varieties that are grown with no shade canopy. These modern plantations require an intensive input of pesticides and chemical fertilizers to sustain yields. Sun coffee also offers little benefit as a wildlife habitat. Studies in Mexico and Colombia found 94% fewer bird species in sun-grown coffee than in shade-grown coffee. Coffee "forests" are significant and cover 2.7 million hectares (more than 6.6 million acres) in Central America, Colombia, and the Caribbean. As of this writing, about 40% of all shade-grown plantations in this region have already been converted to sun.

How can we slow this loss of critical habitat? One way is through the power of the marketplace. Shade-grown coffee has traditionally been the highest-quality coffee available. With the increase in sales of specialty coffees, especially organic coffee, farmers growing shade coffee may receive higher prices for their product — an economic incentive to maintain these critical refuges for migratory birds and other forest species.

Careers in Botany

With a career in *botany* you might find yourself conducting research at a university or botanical garden as a *plant taxonomist* (someone who identifies and classifies plants) or as a specialist in some other aspect of plant science such as *anatomy*, *morphology*, *physiology*, or *genetics*. You might end up working as a *plant breeder* in an international research institute or for a seed company developing new varieties of crops or ornamentals.

If you concentrate in *plant biochemistry* you might find a job conducting research on medicinal compounds, on the nutritional value of plants, or on plants as sources of other useful natural products such as perfumes, resins, tannins, or oils. You can also approach these fields of research by becoming a *chemist* with a specialty in plant compounds.

If your specialty is *economic botany*, your focus is on plants that are useful to people. If you are also interested in the cultural context of plant use, your field is *ethnobotany*.

Careers in Anthropology

You don't necessarily have to approach ethnobotany with the background of a formally trained botanist — *anthropology*, the study of human cultures both past and present, is another entryway into that field of work. Because plants play such an important role in human cultures everywhere, many anthropologists have chosen to concentrate their research on plant use. *Ethnobiology* is a broader term applied to the study of how a culture views and interacts with the living organisms that surround them. Another relevant specialization is *linguistics*, the study of language. Because an ethnobotanist must be fluent in the language of the people she or he is studying, a background in linguistics is extremely useful.

Careers in Ecology

Ecology is the study of how organisms interact with their environment. *Plant ecologists* have different levels of focus depending on their specialization. They may study an individual plant, a plant species, a plant community, or an entire ecosystem. Two areas of research in ecology have particular relevance to the connection between people and plants: *agroecology*, the study of ecological interactions in an agricultural ecosystem, and *ethnoecology*, the study of how the people perceive and interact with their own environment.

Home (again) on the Range

At Arrowhead High School in Hartland, Wisconsin, teacher Greg Bisbee and his biology students are involved with the ecological restoration of a local prairie grassland. Their project aims to restore native grasses to a currently degraded ecosystem. They monitor the composition and development of their growing prairie and, among other projects, conduct research on the collection, storage, and germination of seeds of rare grass species.

These students (some of whom are pictured below, at work on the project) are successfully combining science education with environmental action, in their commitment to the three R's: research, restoration, and recovery of native grasses.

photograph by Greg Bisbee

One very exciting area of ecology that might aid our planet is restoration ecology. A *restoration ecologist* works on the "rehabilitation" of degraded ecosystems. They may work to restore a wetland, clean up a polluted estuary, or turn a landfill into a park. The result will not be identical to the original ecosystem, but every effort is made to restore basic ecological processes such as the cycling of nutrients and water. The related field of *phytoremediation* uses plants to remove, contain, or neutralize environmental contaminants. It is a promising technology that addresses clean-up of heavy metals (such as cadmium and lead), PCBs, explosives (often found at abandoned military sites), and other toxins or pollutants that have leached into the soil.

Volunteer Opportunities

Internships and volunteer positions provide exciting and meaningful opportunities to learn while making a contribution to an organization that interests you. Many institutions such as botanical gardens, conservation organizations, museums, zoos, and parks have formal volunteer or internship programs already in place. Many such organizations post volunteer opportunities on their websites (see Appendix A for a brief list to use as a jumping-off point).

You can also seek out your own volunteer position if you have a specific skill you would like to develop. For instance, if you want to learn more about raising plants, you can volunteer at local nursery. If you are interested in environmental law, consider keeping abreast of legislation that deals with conservation, natural habitats, and endangered species; you may even volunteer to campaign for a political candidate whose views you share. If you are interested in ecological restoration, learn all you can about it and start a project in your community.

In addition to the experience and satisfaction volunteers gain from their work, there is another positive aspect to volunteering: Colleges and employers tend to look favorably upon students who have volunteered time to work for causes that they care about.

If you are interested in becoming a volunteer, the main requirements are enthusiasm, a willingness to learn, and a commitment to following through with any responsibilities you accept. See Appendix A for a list of publications, websites, and organizations that could give you ideas.

Unit 9 Questions for Thought

Name: ______________________________ Date: ________________

1. Look over the list of skills and requirements below, on the right. Then, in the space next to each of the occupations listed on the left, write the letters of all skills and requirements that apply to that job:

________ Horticulturalist
________ Naturalist
________ Biological illustrator
________ Environmental lawyer
________ Economist
________ Plant taxonomist
________ Biochemist
________ Ethnoecology
________ Phytoremediation specialist
________ Ecological restoration specialist
________ Forestry worker
________ Musuem curator
________ Horticultural therapist
________ Alternative marketer
________ Plant geneticist

A = artistic abilities
B = business skills
C = college degree (four years or more)
I = working mostly inside
M = mechanical abilities
O = working mostly outside
P = people skills
S = scientific abilities
T = mathematical abilities
U = computer skills
W = writing abilities

2. Using the letter codes from question 1, list the skills and interests that appeal to you the most and that represent your strengths: ______________________________

3. Which of the careers listed above most closely match your own skills and interests?

____

4. Which of the careers discussed in this chapter is most appealing to you and why?

____

5. Describe a cultural ritual involving plants that you have seen, and explain the meaning of the ritual. ______________________________

____

____

____

6. When planning a garden, what are some things that can help make it successful?

____

____

7. What are some volunteer activities you could engage in that could contribute to the science of ethnobotany or conservation of natural resources? ______________________________

____

____

Unit 9 Activities

Creating Plant-Based Gifts

A. Pressed Flower Displays

Pressed flowers can be used to decorate a bookmark or a greeting card, or can simply be put on display by themselves.

Materials needed
Flowers
Attractive heavy paper
Glue
Scissors or a paper cutter
Clear contact paper or a laminating machine (optional)

Directions

1. Press some attractive flowers using methods described in the Unit 3 section on making herbarium specimens.

2. Cut a piece of your good paper to the size that is appropriate for the object you want to make (for instance, a bookmark that is 1" × 5", a greeting card that is 5" × 4" when folded in half, or a display for framing that is 8" × 10").

3. Arrange the pressed flowers on good paper. Once you are pleased with the arrangement of them, carefully glue them into place.

4. If you would like to laminate your work (highly recommended for a bookmark in particular), place it between two sheets of clear contact paper or laminate it in a laminating machine.

B. Floral Stationery

This stationery uses pressed flowers added to stationery from a store or incorporated into homemade paper.

Materials needed
Flowers
Stationery
Glue

Directions

1. Press some attractive flowers using methods described in the Unit 3 section on making herbarium specimens.

2. If you are using store-bought stationery, go to step 4 below.

3. To make your own stationery, follow the directions for papermaking in Unit 6, with this addition: At step 5 of that activity, before placing the blotter over the pulp, position the pressed flowers on the surface of the pulp and gently press them into the pulp so they are still visible. Continue the activity as described. If you would also like to add pressed flowers to the paper after it is dry, go to step 4 below.

4. Arrange the pressed flowers on good paper. Once you are pleased with the arrangement of them, carefully glue them into place.

C. Herbal Shampoo

Different herbs impart different benefits to the hair. For instance, rosemary stimulates the scalp. Chamomile brings out the natural highlights. Peppermint and lavender are good for oily hair. Tea tree in shampoo can help reduce dandruff problems. Other popular plants to include in shampoo are coconut, lemon, and aloe vera. You can use the following basic shampoo recipe and add whatever herbs you like to it.

Materials needed

One quart of water or herbal tea
4 ounces castille soap flakes
8 drops of essential oil
Pot, mixing bowl, and spoon
Small glass or plastic containers with lids and labels

Directions

1. Boil the water or herbal tea and pour it over the soap flakes in a bowl. Stir until the soap dissolves.

2. Add the essential oil if you are not already using an herbal tea.

3. Pour the shampoo into small containers and label them. (Make sure to write a warning on the label: "Not for internal use.")

Appendix A
Sources and Resources

This appendix provides supplemental information about books, web sites, and organizations you can turn to for more information about ethnobotany, economic botany, conservation, ecology, and other topics discussed in this book. Annotations are given for resources that aren't evident just by their names alone.

Books

Biological Diversity and Conservation

Ausubel, Kenny. 1994. *Seeds of Change, The Living Treasure.* New York: Harper Collins. — Tells the story of the growing movement to restore biodiversity and revolutionize the way we think about food.

Crosby, Alfred W. 1972. *The Columbian Exchange: Biological and Cultural Consequences of 1492.* Westport, CT: Greenwood. — The impact of the exchange of biological resources between the Old World and the New World after the arrival of Columbus.

Primack, Richard B. 1995. *A Primer of Conservation Biology*. Sunderland, MA: Sinauer Associates.

Samson, Fred B., and Fritz L. Knopf, editors. 1996. *Prairie Conservation: Preserving North America's Most Endangered Ecosystem*. Washington, DC: Island Press.

Tuxill, John. 1999. Appreciating the Benefits of Plant Biodiversity. In *State of the World 1999: A Worldwatch Institute Report on Progress toward a Sustainable Society*. New York: W. W. Norton. — Information on plants threatened with extinction.

Wilson, Edward O. 1993. *The Diversity of Life*. New York: W. W. Norton. (Looks at biodiversity from an evolutionary perspective).

Wilson, Edward O., editor. 1988. *Biodiversity*. Washington, DC: National Academy Press. (A series of articles about biodiversity and why we are losing it, how to preserve it, human dependence on it, its value, and restoration ecology).

Botany

Bonnet, Robert L., and G. Daniel Keen. 1989. *Botany: 49 Science Fair Projects*. New York: McGraw-Hill.

Gerard, John. 1975. *The Herbal or General History of Plants: The Complete 1633 Edition as Revised and Enlarged by Thomas Jefferson.* New York: Dover Publications.

Gleason, Henry A., and Arthur Cronquist. 1991. *Manual of Vascular Plants of Northeastern United States and Adjacent Canada.* 2d edition. Bronx: New York Botanical Garden.

Jain, Sudhanshu Kumar, and R. R. Rao. 1977. *A handbook of field and herbarium methods.* New Delhi, India: Today & Tomorrow's Printers and Publishers.

Moldenke, Harold N., and Alma Moldenke. 1952. *Plants of the Bible.* New York: Dover.

Quattrocchi, Umberto. 1999. *CRC World Dictionary of Plant Names: Common Names, Scientific Names, Eponyms, Synonyms, and Etymology.* Boca Raton, FL: CRC Press.

Raven, Peter H., Ray F. Evert, and Susan E. Eichhorn. 1986. *Biology of Plants.* New York: Worth Publishers. — General botany; 775 pages.

Ecology

Buchmann, Stephen L., and Gary Paul Nabhan. 1996. *The Forgotten Pollinators.* Washington, DC: Island Press. — The importance of pollinators and their conservation.

Calabi, Prassede. 1996. *Ecology, A Systems Approach.* Dubuque, IA: Kendall/Hunt.

Colfer, Carol J. Pierce. 1997. *Beyond Slash and Burn: Building on Indigenous Management of Borneo's Tropical Rain Forests.* Advances in Economic Botany volume 11. Bronx: New York Botanical Garden.

Harker, Donald F. 1995. *Where We Live: A Citizen's Guide to Conducting a Community Environmental Inventory.* Washington, DC: Island Press.

Prance, Ghillean T., editor. 1986. *Tropical Rain Forests and the World Atmosphere.* Boulder, CO: Westview Press.

Randall, John M., and Janet Marinelli, editors. 1996. *Invasive Plants: Weeds of the Global Garden.* Brooklyn: Brooklyn Botanic Garden.

Wackernagel, Mathis, and William Rees. 1996. *Our Ecological Footprint: Reducing Human Impact on the Earth.* Gabriola Island, British Columbia: New Society Publishers.

Environmental Studies and Issues

Akbari, Hashem, et al., editors. 1992. *Cooling Our Communities: A Guidebook on Tree Planting and Light-colored Surfacing.* Washington, DC: U.S. Environmental Protection Agency, Office of Policy Analysis, Climate Change Division.

Arms, Karen. 2000. *Holt Environmental Science.* Austin, TX: Holt, Rinehart & Winston.

Bormann, F. Herbert, Diana Balmori, and Gordon T. Geballe. 1993. *Redesigning the American Lawn: A Search for Environmental Harmony.* New Haven: Yale University Press. — Environmental impact of intensive use of chemicals and oil to maintain our lawns and suburban green spaces.

Brower, Michael, and the Union of Concerned Scientists. 1999. *The Consumer's Guide to Effective Environmental Choices: Practical Advice from the Union of Concerned Scientists.* New York: Crown Books.

Brown, Lester, editor. 1997. *State of the World 1997. A Worldwatch Institute Report on Progress Toward a Sustainable Society.* New York: W. W. Norton. — Food scarcity trends, preserving global cropland, climate change, and the process of valuing nature's services.

Howe, Jim. 1997. *Balancing Nature and Commerce in Gateway Communities.* Washington, DC: Island Press. — Planning economic development of communities near national parks and public lands to avoid negative impact on those lands' environmental health; case studies.

Nelson, Richard, editor. 1991. *Helping Nature Heal: An Introduction to Environmental Restoration*. Berkeley: Ten Speed Press.

Ponting, C. 1991. *A Green History of the World: The Environment and the Collapse of Great Civilizations.* New York: Penguin Books.

Ethnobotany, General or Introductory

Alexiades, Miguel N. 1996. *Selected Guidelines for Ethnobotanical Research: A Field Manual.* Bronx: New York Botanical Garden Press.

Anderson, Edward F. 1993. *Plants and People of the Golden Triangle: Ethnobotany of the Hill Tribes of Northern Thailand*. Portland, OR: Timber Press.

Balick, Michael & Cox, Paul Alan. 1996. *Plants, People and Culture: The Science of Ethnobotany*. New York: Scientific American Library.

Brush, Stephen B., and Doreen Stabinsky, editors. 1996. *Valuing Local Knowledge: Indigenous People and Intellectual Property Rights.* Washington, DC: Island Press.

Cotton, C. M. 1996. *Ethnobotany: Principles and Applications.* New York: John Wiley.

Duke, James A., and Rodolfo Vasquez. 1994. *Amazonian Ethnobotanical Dictionary.* Boca Raton, FL: CRC Press.

Given, David R., and Warwick Harris. 1994. *Techniques and Methods of Ethnobotany: As an Aid to the Study, Evaluation, Conservation and Sustainable Use of Biodiversity*. London: Commonwealth Secretariat.

Jain, Sudhanshu Kumar. 1991. *Dictionary of Indian Folk Medicine and Ethnobotany: A Reference Manual of Man–Plant Relationships, Ethnic Groups, and Ethnobotanists in India*. New Delhi, India: Deep Publications.

Jain, Sudhanshu Kumar, editor. 1997. *Contribution to Indian Ethnobotany*. 3rd edition. Jodhpur, India: Scientific Publishers.

Jain, Sudhanshu Kumar, and M. Mudgal. 1999. *A Handbook of Ethnobotany.* Dehra Dun, India: Bishen Singh Mahendra Pal Singh.

Johnson, Timothy. 1999. *CRC Ethnobotany Desk Reference*. Boca Raton, FL: CRC Press.

Kuhnlein, Harriet V., and Nancy J. Turner. 1991. *Traditional Plant Foods of Canadian Indigenous Peoples: Nutrition, Botany, and Use.* Philadelphia: Gordon & Breach Science.

Martin, Gary J. 1995. *Ethnobotany: A Methods Manual*. London: Chapman & Hall.

Nabhan, Gary Paul. 1985. *Gathering the Desert.* Tucson: University of Arizona Press. — Ethnobotany of the American Southwest.

Nazarea, Virginia D. 1998. *Cultural Memory and Biodiversity.* Tucson: University of Arizona Press. — Guidelines for collecting and conserving traditional cultural knowledge to complement the agricultural and botanical study of important crops.

Plotkin, Mark. 1994. *Tales of a Shaman's Apprentice: An Ethnobotanist Searches for New Medicines in the Amazon Rain Forest*. New York: Penguin.

Prance, Ghillean T., and Michael J. Balick, editors. 1990. *New Directions in the Study of Plants and People: Research Contributions from the Institute of Economic Botany.* Advances in Economic Botany, vol. 8. Bronx: New York Botanical Garden.

Rea, Amadeo M. 1997. *At the Desert's Green Edge: An Ethnobotany of the Gila River Pima*. Tucson: University of Arizona Press. — Ethnobotany of the Akimel O'odham, or Pima Indians, of the northern Sonoran Desert.

Schultes, Richard Evans, and Robert F. Raffauf. 1992. *Vine of the Soul: Medicine Men, Their Plants and Rituals in the Colombian Amazonia*. Oracle, AZ: Synergetic Press.

Schultes, Richard Evans, and Siri von Reis, editors. 1995. *Ethnobotany, Evolution of a Discipline*. Portland, OR: Dioscorides Press.

Wagner, H., Hiroshi Hikino, and Norman R. Farnsworth, editors. 1985. *Economic and Medicinal Plant Research*. Orlando, FL: Academic Press.

Whittle, Tyler. 1988. *The Plant Hunters*. New York: PAJ Publications. — An examination of collecting with an account of careers and methods of a number of those who have searched the world for wild plants.

Field Guides and Identification Aids

Batson, Wade T. 1977. *A Guide to the Genera of Native and Commonly Introduced Ferns and Seed Plants of Eastern North America, from the Atlantic to the Great Plains, from Key West-southern Texas into the Arctic.* 3d edition. New York: John Wiley.

Baumgardt, John Philip. 1982. *How to Identify Flowering Plant Families: A Practical Guide for Horticulturists and Plant Lovers.* Portland, OR: Timber Press.

Bell, C. Ritchie, and Bryan Taylor. 1982. *Florida Wild Flowers and Roadside Plants.* Chapel Hill, NC: Laurel Hill Press.

Brill, Steven, with Evelyn Dean. 1994. *Identifying and Harvesting Edible and Medicinal Plants in Wild (and Not So Wild) Places.* New York: Hearst Books.

Brown, Lauren. 1976. *Wildflowers and Winter Weeds.* New York: W. W. Norton. — Illustrated guide, with identification key, to more than 135 plants as they appear during winter in the northeastern United States.

Chadde, Steve W. 1998. *A Great Lakes Wetland Flora: A Complete, Illustrated Guide to the Aquatic and Wetland Plants of the Upper Midwest.* Calumet, MI: PocketFlora Press.

Choukas-Bradley, Melanie. 1987. *City of Trees: The Complete Field Guide to the Trees of Washington, D.C.* Revised edition. Baltimore: Johns Hopkins University Press.

Clark, Lynn G., editor. 1996. *Agnes Chase's First Book of Grasses: The Structure of Grasses Explained for Beginners.* 4th edition, revised. Washington, DC: Smithsonian Institution Press. — An update of the classic guide to identifying grasses by becoming familiar with their structure.

Clovis, Jesse F., et al. 1972. *Common Vascular Plants of the Mid-Appalachian Region.* Morgantown, WV: Book Exchange.

Cronquist, Arthur. 1979. *How to Know the Seed Plants.* Dubuque, IA: Wm. C. Brown.

Dorn, Robert D. 1988. *Vascular Plants of Wyoming.* Cheyenne, WY: Mountain West Publications.

Elias, Thomas S., and Peter A. Dykeman. 1990. *Edible Wild Plants: A North American Field Guide.* New York: Sterling Publishing.

Farnsworth, Beth. 1991. *A Guide to Trails of Guánica: State Forest and Biosphere Reserve.* Guánica, PR: M. Canals.

Fitter, Richard, and Alastair Fitter. 1984. *Collins Guide to the Grasses, Sedges, Rushes, and Ferns of Britain and Northern Europe.* London: Collins.

Foster, Steven, and James Duke. 1998. *A Field Guide to Medicinal Plants: Eastern and Central North America.* Expanded edition. Peterson Field Guides. Boston: Houghton Mifflin.

Godfrey, Michael A. 1997. *Field Guide to the Piedmont: The Natural Habitats of America's Most Lived-in Region, from New York City to Montgomery, Alabama.* Chapel Hill: University of North Carolina Press.

Harris, James G., and Melinda Woolf Harris. 1994. *Plant Identification Terminology: An Illustrated Glossary*. Spring Lake, UT: Spring Lake Publishing.

Hickman, James C., editor. 1993. *The Jepson Manual: Higher Plants of California.* Berkeley: University of California Press.

Hultman, G. Eric. 1978. *Trees, Shrubs, and Flowers of the Midwest*. Chicago: Contemporary Books.

Jacobson, Arthur Lee. 1989. *Trees of Seattle: The Complete Tree-finder's Guide to the City's 740 Varieties.* Seattle: Sasquatch Books.

Jones, George Neville. 1963. *Flora of Illinois, Containing Keys for Identification of Flowering Plants and Ferns.* 3d edition. Notre Dame: University of Notre Dame Press.

Jaques, H. E. 1958. *How to Know the Economic Plants: An Illustrated Key for Identifying the Plants Used by Man for Food and in Other Personal Ways, with Some Essential Facts about Each Plant.* Dubuque, IA: Wm. C. Brown.

Kozloff, Eugene N., and Linda H. Beidleman. 1994. *Plants of the San Francisco Bay Region, Mendocino to Monterey.* Pacific Grove, CA: Sagen Press.

Lawrence, Susannah. 1984. *The Audubon Society Field Guide to the Natural Places of the Mid-Atlantic States.* New York: Pantheon Books.

Mader, Ron. 1998. *Mexico: Adventures in Nature*. Santa Fe, NM: John Muir Publications.

Mahler, Richard. 1997. *Guatemala: Adventures in Nature*. Santa Fe, NM: John Muir Publications.

Mahler, Richard, and Steele Wotkyns. 1997. *Belize: Adventures in Nature*. Santa Fe, NM: John Muir Publications.

Mohlenbrock, Robert H. 1984. *The Field Guide to U.S. National Forests.* New York: Congdon & Weed.

Niering, William A. 1997. *National Audubon Society Nature Guide to Wetlands*. New York: Alfred A. Knopf.

Petrides, George A., and Roger Tory Peterson. 1998. *A Field Guide to Western Trees*. 2d edition. Peterson Field Guides. Boston: Houghton Mifflin.

Petrides, George A., Janet Wehr, and Roger Tory Peterson. 1998. *A Field Guide to Eastern Trees*. Revised edition. Peterson Field Guides. Boston: Houghton Mifflin.

Schofield, Eileen K. 1984. *Plants of the Galápagos Islands.* New York: Universe Books.

Smith, Edwin B. 1994. *Keys to the Flora of Arkansas.* Fayetteville: University of Arkansas Press.

Strange Sheck, Ree. 1998. *Costa Rica: Adventures in Nature*. Santa Fe, NM: John Muir Publications.

Symonds, George. 1973. *Shrub Identification Book.* New York: William Morrow. — Visual identification guide to shrubs, vines, and groundcovers, with more than 3,500 photographs.

Welsh, Stanley L., Michael Treshow, and Glen Moore. 1965. *Common Utah Plants*. 2d edition. Provo, UT: Brigham Young University Press.

Wiggers, Raymond. 1994. *The Plant Explorer's Guide to New England*. Missoula, MT: Mountain Press.

Wunderlin, Richard P. 1982. *Guide to the Vascular Plants of Central Florida.* Tampa: University Presses of Florida.

Food and Nutrition

Margen, Sheldon, and the Editors of the University of California at Berkeley Wellness Letter. 1992. *The Wellness Encylcopedia of Food and Nutrition: How to Buy, Store, and Prepare Every Fresh Food.* New York: Random House.

Duyff, Roberta Larson. 1998. *The American Dietetic Association's Complete Food and Nutrition Guide.* Minneapolis: Chronimed Publishing.

Tamborlane, William V. (editor in chief), Janet Z. Weiswasser (managing editor), Teresa Fung, Nancy A. Held, and Tara Prather Liskov (editors). 1997. *The Yale Guide to Children's Nutrition.* New Haven: Yale University Press.

Gardening and Agriculture

Cave, Philip. 1993. *Creating Japanese Gardens.* Boston: Charles E. Tuttle.

Crockett, James Underwood. 1977. *Crockett's Victory Garden.* Boston: Little, Brown. — Clear, seasonal advice on how to start a garden, by the host of the PBS television series.

Dierks, Leslie. 1994. *Wreaths from the Garden: 75 Fresh and Dried Floral Wreaths to Make.* New York: Sterling Publications.

Engel, David H. 1986. *Creating a Chinese Garden.* Portland, OR: Timber Press.

Malcolm Hillier, editor. 1990. *Flower Arranging.* Pleasantville, NY: Reader's Digest Association.

Kohnke, Helmut & Franzmeier, D. P. *Soil Science Simplified, Fourth Edition.* Illinois: Waveland Press, 1980.

Levitan, Lois. 1980. *Improve Your Gardening with Backyard Research.* Emmaus, PA: Rodale Press.

Marinelli, Janet, editor. 1994. *Going Native: Biodiversity on Our Own Backyards.* Handbook #140. Brooklyn: Brooklyn Botanical Garden.

Packer, Jane. 1995. *The Complete Guide to Flower Arranging.* New York: Dorling Kindersley.

Schneck, Marcus H. 1994. *Creating a Hummingbird Garden: A Guide to Attracting and Identifying Hummingbird Visitors.* New York: Simon & Schuster.

Schneck, Marcus H. 1994. *Creating a Butterfly Garden: A Guide to Attracting and Identifying Butterfly Visitors.* New York: Simon & Schuster.

General Biology

Brusca, Johnson. 1994. *Biology, Visualizing Life.* Austin, TX: Holt, Rinehart and Winston.

General Science

Barhydt, Frances Bartlett & Morgan, Paul W. 1993. *The Science Teacher's Book of Lists.* Englewood Cliffs, NJ.: Prentice Hall. — Lists of plant parts, endangered species, plant products, metrics, simple statistics, and useful formulas.

Edwards, Gabrielle I. & Cimmino, Marion. 1994. *Laboratory Techniques for High Schools, A Workbook of Biomedical Methods*. Hauppauge, NY: Barron's Educational Series.

Smith, Norman F. 1990. *How to Do Successful Science Projects.* Revised edition. New York: Simon & Schuster.

Van Cleave, Janice Pratt. 1993. *A + Projects in Biology: Winning Experiments for Science Fairs and Extra Credit*. New York: John Wiley.

MEDICINAL AND USEFUL PLANTS

Arvigo, Rosita & Balick, Michael. *Rainforest Remedies, One Hundred Healing Herbs of Belize.* Twin Lakes, WI: Lotus Press, 1993.

Balick, Michael J., Elaine Elisabetsky, and Sarah A. Laird. 1996. *Medicinal Resources of the Tropical Forest: Biodiversity and Its Importance to Human Health.* New York: Columbia University Press.

Beckstrom-Sternberg, Steven M., and James A. Duke. 1996. *CRC Handbook of Medicinal Mints (Aromathematics): Phytochemicals and Biological Activities.* Boca Raton, FL: CRC Press.

Bliss, Anne. 1979. *North American Dye Plants.* New York: Scribner.

Blumenthal, Mark, editor. 1998. *Commission E Monographs.* Austin, TX: American Botanical Council; Boston: Integrative Medicine Communications. — The most accurate information in the world on the safety and efficacy of herbs and phytomedicines; technical and detailed.

Bremness, Lesley. 1994. *Herbs, The Visual Guide to More Than 700 Herb Species From Around the World.* London: Dorling Kindersley.

Chevallier, Andrew. 1996. *The encyclopedia of Medicinal Plants: A Practical Reference Guide to More than 500 Key Medicinal Plants and Their Uses.* New York: Dorling Kindersley.

Dobelis, Inge N., editor. 1986. *Magic and Medicine of Plants.* Pleasantville, NY: Reader's Digest. — History and list of useful plants, and how to grow and use herbs.

Estévez M., Arcenio, and Freddy Báez. 1992. *Plantas curativas: Usos populares y científicos.* Santo Domingo: Instituto de Medicina Natural.

Fern, Ken. 1997. *Plants for a Future: Edible and Useful Plants for a Healthier World.* Clanfield, England: Permanent Publications; distributed in the United States by Rodale Publications, Kutztown, PA.

Fussell, Betty Harper. 1992. *The Story of Corn.* New York: Knopf.

Griggs, Barbara. 1991. *The Green Pharmacy, The History and Evolution of Western Herb Medicine.* Rochester, VT: Healing Arts Press.

Grainger, Janette & Moore, Connie. 1995. *Natural Insect Repellents for Pets, People and Plants.* Austin, TX: Herb Bar.

Heiser, Charles B., Jr. 1990. *Seed to Civilization: The Story of Food.* New edition. Cambridge: Harvard University Press.

Hobhouse, Henry. 1985. *Seeds of Change: Five Plants that Transformed Mankind.* New York: Harper & Row.

Jain, Sudhanshu Kumar. 1975. *Medicinal Plants.* 2d revised edition. New Delhi, India: National Book Trust.

Jain, Sudhanshu Kumar, and Robert A. DeFilipps. 1991. *Medicinal Plants of India*. Algonac, MI: Reference Publications.

Jain, Sudhanshu Kumar, B. K. Sinha, and R. C. Gupta. 1991. *Notable Plants in Ethnomedicine of India*. New Delhi, India: Deep Publications.

Kindscher, Kelly. 1992. *Medicinal Wild Plants of the Prairie: An Ethnobotanical Guide*. Lawrence, KS: University Press of Kansas.

Lewington, Anna. 1990. *Plants for People*. New York: Oxford University Press.

Mabey, Richard. 1978. *Plantcraft: A Guide to the Everyday Use of Wild Plants*. New York: Universe Books.

Martin, Marie A. 1971. *Introduction à l'ethnobotanique du Cambodge*. Paris: Éditions du Centre National de la Recherche Scientifique.

Mitchell, Faith. 1999. *Hoodoo Medicine: Gullah Herbal Remedies*. Revised edition. Columbia, SC: Summerhouse Press. — Herbal medicines used from the 1600s until recent decades by African slaves and their freed descendants, who settled in the South Carolina Sea Islands.

Ody, Penelope. 1993. *The Complete Medicinal Herbal*. London: Dorling Kindersley.

Patnaik, Naveen. 1993. *The Garden of Life, an Introduction to the Healing Plants of India*. New York: Doubleday.

Piper, Jacqueline M. 1992. *Bamboo and Rattan: Traditional Uses and Beliefs*. New York: Oxford University Press.

Polunin, Miriam. 1997. *Healing Foods: a Practical Guide to Key Foods for Good Health*. New York: Dorling Kindersley.

Potterman, David. 1983. *Culpeper's Color Herbal*. New York: Sterling Publishing.

Prance, Ghillean T., Derek J. Chadwick, and Joan Marsh, editors. 1994. *Ethnobotany and the Search for New Drugs*. New York: John Wiley.

Prance, Ghillean T. 1985. *Leaves: The Formation, Characteristics, and Uses of Hundreds of Leaves Found in All Parts of the World*. New York: Crown.

Schultz, Kathleen. 1975. *Create Your Own Natural Dyes*. New York: Sterling Publishing.

Sheldon, Jennie Wood, Michael Balick, and Sarah Laird. 1997. *Medicinal Plants: Can Utilization and Conservation Coexist?* Bronx: New York Botanical Garden Press.

Simpson, Beryl Britnell, and Molly Conner Ogorzaly. 1995. *Economic Botany: Plants in Our World*. 2d edition. New York: McGraw-Hill. — Excellent general resource on economic plants.

Vickery, Margaret L., and Brian Vickery. 1979. *Plant Products of Tropical Africa*. New York: Macmillan.

Wagner, H., Hiroshi Hikino, and Norman R. Farnsworth, editors. 1985– . *Economic and Medicinal Plant Research*. Orlando, FL: Academic Press.

Wilson, Roberta. 1995. *Aromatherapy for Vibrant Health and Beauty*. Garden City Park, NY: Avery Publishing Group.

Wolfson, Evelyn. 1993. *From the Earth to Beyond the Sky: Native American Medicine*. New York: Houghton Mifflin.

Wyman, Leland C., and Stuart K. Harris. 1941. *Navajo Indian Medical Ethnobotany*. Albuquerque, NM: University of New Mexico Press.

Journals, Magazines, and Newsletters

Ambio: A Journal of the Human Environment. Published by the Royal Swedish Academy of Sciences, its aim is to report and analyze developments in environmental research, policy and related activities. Contact: Royal Swedish Academy of Sciences, Box 50005, S-104 05 Stockholm, Sweden; Tel. +46.8.6739500 or 673154744, Fax +46.8.673166251

Biodiversity and Conservation. Published by Kluwer Academic, it presents papers dealing with the description, analysis, and conservation of biodiversity, and economic, social, political and practical management issues. Contact: Kluwer Academic Publishers, Journals Department, P.O. Box 322, 3300 AH Dordrecht, The Netherlands; Tel: (+31) 78 639 23 92; Fax: (+31) 78 654 64 74; email: services@wkap.nl

Conservation Biology. This journal of the Society for Conservation Biology includes research papers and essays on conservation and natural resource issues, as well as book reviews and a section on international conservation news. Contact: Journals Department, Blackwell Science, Commerce Place, 350 Main Street, Malden, MA 02148-5018; phone: 781.388.8250; fax: 781.388.8255; email: csjournals@blacksci.com

Economic Botany. This journal of the Society for Economic Botany includes original research papers on past, present, and potential uses of plants, and documents the rich relationship between plants and people around the world. Contact: The New York Botanical Garden Press, 200th St. & Kazimiroff Blvd., Bronx, NY 10458; phone: 718.817.8721; email: nybgpress@nybg.org; web site: www.nybg.org/bsci/spub/catl/CATAL2.html#ECO

Ethnobotanical Catalog of Seeds. This commercial catalog costs $1 and is a mail-order source of high-quality seeds for herbs, trees, vegetables, and some ornamentals. Contact: J. L. Hudson Seedsman, P.O. Box 1058, Redwood City, CA 94064

Grupo Ethnobotanico LatinoAmericano. A publication in Spanish which focuses on the ethnobotany of Latin America and the Caribbean. Contact: Javier Caballero, Jardin Botanico, Universidad Nacional Autonoma de Mexico, Apartado Postal 70-614, Mexico D.F. 04510, Mexico; fax: +52.5.6229046; email: jcnieto@servidor.unam.mx.

The Herb Quarterly. This commercial magazine covers the history and folklore of herbs, gardening and cooking with herbs, crafts, and medicinal uses. Contact: P.O. Box 689, San Anselmo, CA 94979-0689; phone: 800.371.4372; web site: www.herbquarterly.com

HerbalGram. This journal of the American Botanical Council and the Herb Research Foundation aims to educate the public on the use of herbs and phytomedicines. Contact: The American Botanical Council, P.O. Box 201660, Austin, TX 78720-1660; fax: 512.331.1924; email: custserv@herbalgram.org; web site: www.herbalgram.org/herbalgram/index.html

Indigenous Knowledge and Development Monitor. Produced in close collaboration with 22 international, regional, and national Indigenous Knowledge Resource Centres, this newsletter contains articles describing the contribution of indigenous knowledge to the process of sustainable development in many parts of the world. Also includes information on resource centers, networks, research, conferences, databases, publications, and films. Contact: Centre for International Research and Advisory Networks (CIRAN/Nuffic), P.O. Box 29 777, 2502 LT The Hague, The Netherlands; phone: 31.70.4260321; fax: 31.70.4260329; email: tick@nufficcs.nl; web site: www.nuffic.nl/ciran/ikdm/index.html

Journal of Ethnobiology. Published by the Society of Ethnobiology, this journal features scientific articles that capture the range and impact of contemporary ethnobiological research. Contact: Society of Ethnobiology, Missouri Botanical Garden, Center for Plant Conservation, P.O. Box 299, St. Louis, MO 63166. Phone: 314.577.9450

Journal of Herbal Pharmacotherapy: Innovations in Clinical and Applied Evidence-Based Herbal Medicinals. email: getinfo@haworthpressinc.com

Journal of Herbs, Spices, and Medicinal Plants. email: getinfo@haworthpressinc.com

Journal of Nutraceuticals, Functional, and Medical Foods: Product Development, Commercialization, and Policy Issues. email: getinfo@haworthpressinc.com

Organic Gardening. A Rodale Press Publication. 33E. Minor Street. Emmaus, PA 18098. Phone 610.967.5171.

Plants and People Handbook. This newsletter provides information on sustainable and equitable use of plant resources.

PhytoPharmica. Prescription for Health, a newsletter for health care professionals interested in natural health care alternatives. 825 Challenger Drive, Green Bay, WI 54311

Laboratory and Biological Supplies

The tests in Unit 4 are loosely based on guidelines that come with the Lab-Aids Food Nutrient Analysis Kit, but they are appropriate for use with lab materials from other biological supply manufacturers. Kits and other supplies can be purchased from the following biological supply catalogues.

Carolina Biological Supply
800-334-5551
www.carolina.com

Sargent-Welch
800-727-4368
www.sargentwelch.com/index.html

Science Stuff
800-795-7315
sciencestuff.com/menu.htm

VWR Scientific Products
800-932-5000
www.vwrsp.com

Wards Scientific
800-962-2660
www.wardsci.com

Organizations

The Biodiversity Support Program — A consortium of World Wildlife Fund, The Nature Conservancy, and World Resources Institute that promotes the conservation of biodiversity and the improvement of human livelihoods in developing countries. Contact: c/o World Wildlife Fund, 1250 24th St. NW, Washington, DC 20037; phone: 202.861.8347; web site: www.bsponline.org

Conservation International — Conducts conservation initiatives and scientific activities. Contact: 2501 M St. NW, Suite 200, Washington, DC 20037; phone: 202.429.5660; fax: 202.887.0193; web site: www.conservation.org

Cultural Survival — Defends human rights and cultural autonomy of indigenous peoples of Canada. Contact: 96 Mount Auburn St., Cambridge, MA 02138; web site: www.cs.org

EarthWatch Institute — Promotes sustainable conservation of natural resources and cultural heritage by creating partnerships between scientists, educators, and the general public. Volunteers can go on field trips to provide hands-on assistance in ongoing scientific and cultural research projects, many of which have an environmental or ethnobotanical focus. *Earthwatch* is the organization's magazine that highlights these expeditions to many parts of the world. Contact: 680 Mount Auburn St., Watertown, MA 02272-9104; phone: 617.926.8200; email: info@earthwatch.org; web site: www.earthwatch.org

National Wildlife Federation — Works on environmental education, issues, outdoor adventures, and camps. Contact: 8925 Leesburg Pike, Vienna, VA 22184; web site: www.nwf.org/nwf

Native Seeds — A group that grows, distributes, and collects rare and native seeds. Contact: 2509 N. Campbell Ave. No. 32.5, Tucson, AZ 85719; phone: 602.327.9123

The Nature Conservancy — Operates the world's largest private system of nature sanctuaries to protect imperiled species of plants and animals. Contact: 4245 N. Fairfax Dr., Suite 100, Arlington, VA 22203-1606, phone: 703.841.5300; web site: www.tnc.org

The People and Plants Initiative — A collaborative effort between the World Wildlife Fund, The United Nations Educational, Scientific and Cultural Organization (UNESCO), and the Royal Botanical Gardens, Kew. Along with its focus on ethnobotany, conservation, and community development, the group offers publications and ethnobotanical training programs. Contact: Royal Botanic Gardens, Kew, Richmond, Surrey TW9 3AE, UK; phone: +44.181.332.5706; web site: www.rbgkew.org.uk/peopleplants

Rainforest Action Network — Campaigns to protect rainforests, gives news and information. The web site has a kids' corner and links to like-minded organizations. Contact: 221 Pine St., Suite 500, San Francisco, CA 94104, phone: 415.398.4404; email: rainforest@ran.org; web site: www.ran.org/ran/intro.html

Seed Savers Exchange — A group of people who grow and exchange rare and heirloom seed varieties. Contact: 3076 North Winn Rd., Decorah, IA 52101

World Wildlife Fund — Directs its conservation efforts toward three global goals: protecting endangered spaces, saving endangered species, and addressing global threats. Contact: 1250 24th St. NW, Washington, DC 20037; web site: www.wwf.org

Web Sites

Biological Diversity and Conservation

Biodiversity and Biological Collections Web Server — Lists organizations related to biodiversity. Has online images, journals, societies, directories, news and electronic books.
biodiversity.uno.edu

Bioresources Development and Conservation Programme — Shows the connection between biodiversity conservation and socioeconomic development, with articles, projects, and a virtual newsletter.
www.bioresources.org

Center for Applied Biodiversity Science
www.cabs.conservation.org

Center for Biodiversity and Conservation, American Museum of Natural History — Guides to being a "green" consumer, proceedings of biodiversity symposia, publications, newsletters, and more.
research.amnh.org/biodiversity

Convention on Biological Diversity — A partnership among countries that provides for scientific and technical cooperation, access to genetic resources, and the transfer of environmentally sound technologies; multilingual site (available in English, Spanish, and French) includes historical perspective, national reports and official documents on biodiversity, and an information clearinghouse.
www.biodiv.org

Botany and Botanical Gardens

The Arnold Arboretum at Harvard — Living collections of plants and information.
www.arboretum.harvard.edu

California Wildflowers, California Academy of Sciences — Look up wildflowers by color, common name, scientific name, or region; photos of each flower from several different angles, plus information on the plant.
www.calacademy.org/research/botany/wildflow/index.html

Catalog of Botanical Illustration, Smithsonian Institution — 500 illustrations by professional botanical illustrators, for online viewing
www.nmnh.si.edu/botart

Internet Directory for Botany — Very extensive set of links to web sites from or about botanical gardens, botanical organizations, conservation, threatened plants, ethnobotany, economic botany, gardening, botanical images, journals, books, publishers, and much more.
www.helsinki.fi/kmus/botmenu.html

Missouri Botanical Garden
www.mobot.org

National Plant Data Center of the USDA — Links to a plant database and plant related web sites.
trident.ftc.nrcs.usda.gov/npdc

The New York Botanical Garden
www.nybg.org

Plant Trivia Timeline
www.huntington.org/BotanicalDiv/Timeline.html

Royal Botanic Gardens, Kew
www.rbgkew.org.uk/ceb/ebinfo.html

Vascular Plant Image Gallery
www.csdl.tamu.edu/FLORA/gallery.htm

Virtual Foliage, University of Wisconsin–Madison, Botany Dept. — Thousands of digital plant images.
www.wisc.edu/botany/virtual.html

World Wide Web Virtual Library of Botany
www.ou.edu/cas/botany-micro/www-vl

Career and Volunteer Information

U.S. Army Corps of Engineers Volunteer Opportunities — Ideas for volunteering, from giving out visitor information to maintaining nature trails, and from planting seedlings to picking up litter.
www.orn.usace.army.mil/volunteer

The Nature Conservancy, Volunteer and Internship Opportunities — Lists opportunities for every state.
www.tnc.org/involved/html/volopps2.htm

Ecology

Wildland Weeds Management and Research Program, The Nature Conservancy — Information about invasive plants and how they can be controlled and managed.
tncweeds.ucdavis.edu

Economic Botany

Society for Economic Botany (U.S.)
www.econbot.org

The Center for Economic Botany, Royal Botanic Gardens, Kew
www.rbgkew.org.uk

Environmental Studies and Issues

E: The Environmental Magazine — Online version, with links and a marketplace.
www.emagazine.com

Earthshots: Satellite Images of Environmental Change, from the USGS
edcwww.cr.usgs.gov/earthshots/slow/tableofcontents

Envirofacts, Access to Environmental Protection Agency databases
www.epa.gov/enviro/html/first_time.html

The Forest Where We Live — Information and activities about urban forests; includes a "Calculate Your Carbon Debt" applet, a "What You Can Do" section, and case studies of urban forestry projects.
www.lpb.org/programs/forest

Human-Environment Research Laboratory, University of Illinois — Multidisciplinary research laboratory studying the relationships between people and the environments they inhabit, such as urban public housing in Chicago, the urban rural fringe, farms, and highway corridors
www.aces.uiuc.edu/ ~ herl/welcome.html

Sustainable Development — A space for sustainable development organizations to exhibit and share their work.
www.sustainabledevelopment.org

U.S. Army Corps of Engineers Phytoremediation Research
www.wes.army.mil/el/phyto

World Wildlife Fund's *Living Planet Report* online — Quantifies the rate at which nature is disappearing from the earth, and describes how human pressures on natural resources change over time and how these pressures vary from country to country.
www.panda.org/livingplanet/home.shtml

Ethnobotany, General

Dr. Duke's Phytochemical and Ethnobotanical Databases, U.S. Agricultural Research Service — In addition to phytochemical and ethnobotanical databases that can be searched or browsed, there is an ethnobotanical dictionary and links to other databases of interest.
www.ars-grin.gov/duke

EthnobotDB — Worldwide plant uses, searchable by common name, country, family, genus, or uses
probe.nalusda.gov:8300/cgi-bin/browse/ethnobotdb

The People and Plants Initiative
www.rbgkew.org.uk/peopleplants

Indigenous Knowledge and Projects in Specific Geographic Regions

Amazon Conservation Team (ACT) — Information about conservation projects and research that combine indigenous knowledge with Western science to understand, document, and preserve the biological and cultural diversity of the Amazon.
www.ethnobotany.org

American Indian Ethnobotanical Database — Database of thousands of plants used by Native North American Peoples for foods, drugs, dyes, and fibers.
www.umd.umich.edu/cgi-bin/herb

Best practices on Indigenous Knowledge, UNESCO — Explains the meaning of "indigenous knowledge" and "best practices"; provides case studies of indigenous knowledge practices in Africa, Asia, Europe, and Latin America, describing for each the practice, its sustainability, its strengths and weaknesses, and links to the web sites of organization(s) involved.
www.unesco.org/most/bpikpub.htm

Indigenous Knowledge Homepage, The Netherlands Centre for International Research and Advisory Networks — Thorough lists of indigenous knowledge projects, browsable by region, country, topic, or keyword.
www.nuffic.nl/ik-pages/index.html

Indigenous Knowledge Initiative in Africa, from The World Bank — Includes a database of indigenous knowledge in Africa, key resources for more information, and links to participating organizations.
www.worldbank.org/afr/ik/default.htm

Plants of the Machiguenga — Informative and engaging account (with lots of photographs) of a neurologist's search for migraine remedies among the Machiguenga, a native people in eastern Peru, including results of laboratory research he later conducted on Machiguenga plants.
www.montana.com/manu/index.html

Food and Nutrition

BioNutritional Encyclopedia — Provides brief accounts of the benefits of herbs, dietary supplements, minerals, and vitamins; "grades" the scientific support for each claim and includes complete references to the relevant scientific literature.
www.biovalidity.com/gsearch2.cfm

Iowa Soybean Growers Association — Includes information on nutritional values and health benefits of soy foods, plus access to research and scientists who study soybeans.
www.soyfoods.com

Mayo clinic web site on nutrition
www.mayohealth.org

North American Blueberry Council — Includes information on nutritional values and phytochemicals in blueberries.
www.blueberry.org

University of California at Davis, Fruit and Nut Research and Information Center — Information on numerous fruits and nuts, pest management, agricultural information, botanical information, and more.
fruitsandnuts.ucdavis.edu/crops.html

U.S. Food and Drug Administration, Center for Food Safety and Applied Nutrition — Information on dietary supplements, food additives, cosmetics, pesticides, chemical contaminants, and much more.
vm.cfsan.fda.gov/list.html

Gardening and Agriculture

Agripedia, University of Kentucky College of Agriculture — Extremely rich resource for information about agriculture, agricultural economics, forestry, landscape architecture, plant pathology, and plant and soil sciences.
frost.ca.uky.edu/agripedia/index.htm

AgriSurf! The Farmer's Search Engine — Links to more than 16,000 web sites relating to every imaginable aspect of agriculture; organized and searchable.
www.agrisurf.com

The Evolution of Crop Plants — Fascinating online version of a course given by Paul Depts at University of California at Davis. Where did agriculture originate and when? Where did our major crops originate? How were plants modified as a consequence of cultivation? Using lecture notes, print resources, images, and links to relevant web sites, this comprehensive resource tries to answer these and many more questions.
agronomy.ucdavis.edu/GEPTS/pb143/pb143.htm

Garden Web — Resources, discussions, botanical glossary, seed and plant exchange forums, products, and book store.
www.gardenweb.com

National Gardening — Web site of the National Gardening Association, which promotes the benefits of gardening; has literature, grants, a magazine, related links, and curriculum materials.
www.garden.org

Health and Medicine

Healthfinder — User-friendly database with information on nutrition, herbal supplements, and more.
www.healthfinder.gov

Mayo Clinic Health Oasis — Health and nutrition information from an established, highly respected health care and medical research facility.
www.mayohealth.org

Medscape
www.medscape.com

Medicinal Plants, Useful Plants, and Herbal Remedies

Botanical Medicine Information Resources (from the Rosenthal Center for Complementary and Alternative Medicine, Columbia University)
cpmcnet.columbia.edu/dept/rosenthal/Botanicals.html

Herb Research Foundation — News, discussions, and links.
www.herbs.org

The Herbal Bookworm. Reviews herbal books and literature.
www.teleport.com/ ~ jonno

Herbal Encyclopedia. This site has excellent information, but no graphics.
www.wic.net/waltzark/herbenc.htm

University of Washington Medicinal Herb Garden — Images of medicinal plants, with links for each plant to the USDA's Ethnobotanical Database and Medicinal Plants of North America Database, Mrs. Grieve's Modern Herbal, and National Institute of Health's MEDLINE citations via PubMed
www.nnlm.nlm.nih.gov/pnr/uwmhg

Plant Safety

The following two web sites should be consulted to ensure that the plant you wish to study is not dangerous. Consulting these two resources may help you avoid preventable problems:

Cornell University Poisonous Plants Page
www.ansci.cornell.edu/plants/

Canadian Poison Plants Information System
res.agr.ca/brd/poisonpl/title.html

Research Methods

Electrophoresis Chamber Instructions, from the Genetic Science Learning Center at the University of Utah — Building your own gel electrophoresis chamber can be easy, inexpensive, and fun; includes a section on "kitchen electrophoresis experiments," mostly on genetics.
gslc.genetics.utah.edu/basic/units/electrophoresis/index.html

Research Methods Knowledge Base, by William M. Trochim — Includes discussions of foundations of research methods, sampling, measurement, experiment design, data analysis, and write-up of results.
trochim.human.cornell.edu/kb/contents.htm

Statistical Methods

Introductory Statistics, by David Stockburger
www.psychstat.smsu.edu/Introbook/sbk00.htm

Introductory Statistics Tutorial from NASA's Institute on Climate and Planets
icp.giss.nasa.gov/education/statistics

Graphical Techniques (from StatSoft's Statistics Homepage)
www.statsoft.com/textbook/stgraph.html

Appendix B
Fifty-four Useful and Interesting Plants

The following list of 54 useful plants from around the world is arranged by region of origin. Each plant is accompanied by a very brief summary of some of its traditional and current uses, not all of which have been scientifically proven.

The purpose of this list is twofold: First, if you are having difficulty choosing a plant to study, you may turn to this list for help; second, this list is another chance to share with you how important and amazing the plant kingdom can be.

Selecting only 54 plants from around the world to feature in this appendix was not easy. There are many thousands of fascinating and useful plants to choose from. Each plant on this list meets the following criteria:

1. It is usually possible to find or purchase this plant in the United States.
2. It is a legal plant in the United States.
3. It is a relatively safe plant to use.
4. It has a variety of uses.
5. It is not an endangered or rare plant.

Some of the plants I have chosen are naturalized in the United States although they may have originated elsewhere, such as kudzu vine (*Pueraria montana* var. *lobata*) and purple loosestrife (*Lythrum salicaria*). If you can find some new uses for these successful weeds that are now so abundant, you will be making a significant contribution to our environment.

Africa

ALOE • *Aloe vera*
A succulent plant. Gel from leaves helps heal cuts and reduces infection risk. Laxative, emollient, cathartic. Regenerates damaged skin. Used in salves, skin lotion, and shampoo.

COFFEE • *Coffea arabica*
Evergreen shrub. Berries used as a stimulant. Increases alertness, perception. Treats headaches, increases blood flow, heart rate. Excessive use causes shaking, insomnia.

MYRRH • *Commiphora* species
Spiny tree. Gum resin soothes sore throats, stimulant, antiseptic, astringent, expectorant, antispasmodic, antimicrobial, anti-inflammatory. Used as incense.

This list is solely for informational purposes.
Do not attempt to medicate yourself without the supervision of a qualified physician.

Senna • *Cassia senna*

Bean family shrub. Used as a stimulant, cathartic. Commonly used in commercial laxatives. Softens stool, makes bowels move. Eases constipation and nausea.

Tamarind • *Tamarindus indica*

Evergreen tree of the bean family. Edible fruit with B & C vitamins. Helps digestion, stops gas, soothes sore throat, improves appetite, relieves constipation. A mild laxative. Reduces mucus. Good cold remedy.

Central America and the Caribbean

Allspice • *Pimenta officinalis*

Fragrant evergreen tree. Berries have vitamins A, C, B complex, minerals, oil and protein. For sauces and condiments. Digestive stimulant, relieves flatulence, diarrhea, indigestion. Antiseptic and a stimulant.

Avocado • *Persea americana*

Tropical and semitropical tree. Fruit is edible, high in oil. Peel expels head lice and intestinal worms, is a carminative, astringent, relieves coughs. Fruit soothes skin, helps heal wounds, scars, blemishes.

Corn • *Zea mays*

Annual of grass family. Important human and animal food. Corn silk lowers blood pressure, stimulates secretion of bile. A demulcent, diuretic. Reduces kidney stones. Iron, potassium source. Steadies blood sugar.

Papaya • *Carica papaya*

A small, fast growing tropical tree. Fruits used for eating, juice, as a digestive aid, meat tenderizer, laxative. Contains digestive enzymes. Seeds expel worms. High in vitamin A.

Yam • *Dioscorea villosa*

Herbaceous vine with starchy edible tuber. Contains natural steroids and estrogen; relieves menstrual, labor pain and menopausal symptoms. An anti-inflammatory and diuretic.It eases arthritis and cramps.

North America

Goldenseal • *Hydrastis canadensis*

An overcollected woodland herbaceous perennial. Antiseptic, hemoseptic, tonic, laxative, anti-inflammatory, astringent, antibacterial, antifungal. Treats genitourinary disorders.

This list is solely for informational purposes.
Do not attempt to medicate yourself without the supervision of a qualified physician.

PURPLE CONEFLOWER • *Echinacea purpurea*
An herbaceous perennial. An immune system stimulant, anti-inflammatory, antibiotic, detoxifying, antiallergenic, heals wounds, promotes sweating. Indians used for snakebite and swollen glands.

SUNFLOWER • *Helianthus annuus*
A tall annual. Edible seeds have vitamin E and linoleic acid to reduce angina, lower cholesterol and discourage blood clotting in arteries. Antioxidant and anticancer agent. Reduces risk of damage from excess exercise.

WITCH HAZEL • *Hamamelis virginiana*
Deciduous shrub. An astringent, decoction for excess menstrual flow. Treats hemorrhoids, itching. A refreshing aftershave. Helps eliminate eye irritation.

YUCCA • *Yucca filamentosa*
A desert plant with pointed leathery leaves. Used for salves, poultices, to stop bleeding, hair tonic. Antiarthritic. Strong fiber used for weaving and sewing. Relieves skin disorders and sprain injury.

South America

AMARANTH • *Amaranthus caudatus*
An annual. Important food source for Aztecs; grain is high in protein. Has red dye, is an astringent, helps ease menstrual cramps and bleeding. Also used for diarrhea, dysentery, and canker sores.

CAYENNE PEPPER • *Capsicum* species
Annual used as hot spice for food, insecticide, a tonic, carminative, digestive stimulant, eases joint pain and congestion, induces sweating.

CINCHONA • *Cinchona* species
Large tropical tree contains quinine. Was source of first effective antimalarial medicine. Quinine in bark is a tonic, an appetite stimulant, antispasmodic, antibacterial, antiprotozoal. Lowers fever.

NASTURTIUM • *Nasturtium officinale*
A creeping vine. Edible flowers, leaves, seeds that are good in salads. Antibiotic, disinfectant, clears congestion. Seeds are a purgative. A good eyewash.

PINEAPPLE • *Ananas comosus*
A bromeliad plant. Contains bromelain, a digestive enzyme. Edible fruit has fiber, vitamins A and C. Enhances appetite, eliminates gas. Digestive tonic, diuretic.

This list is solely for informational purposes.
Do not attempt to medicate yourself without the supervision of a qualified physician.

POTATO • *Solanum tuberosum*

A perennial with edible tubers that contain starch, vitamins A, B1, B2, C, K and potassium. Juice relieves peptic ulcers and headache, soothes painful joints, rashes, hemorrhoids. Important food.

SARSAPARILLA • *Smilax* species

Woody climbing perennial. Flavor for root beer. Anti-inflammatory, eases skin conditions like eczema. Has natural reproductive hormones to ease symptoms of PMS and enhances virility.

South East Asia, Australia, and South Pacific

BANANA • *Musa paradisiaca*

Potassium source. Relieves constipation, diarrhea. Soothes mucous membranes. Edible fruit. Fibers in leaf bases can be made into strong paper.

CLOVE • *Syzygium aromaticum*

An evergreen tree. Used as an antidote to pain, bad breath, nausea. An antiseptic, stimulant, carminative, analgesic. Eliminates parasites, prevents vomiting. Antibacterial and antifungal, treatment for skin conditions.

EUCALYPTUS • *Eucalyptus globulus*

Used as an antiseptic, expectorant, insect repellant and vapor rub for colds. Clears congestion and eases rheumatic aches and pains.

JAMBUL • *Syzygium cumini*

Evergreen tree. Edible fruit that lowers blood glucose levels, helpful in controlling diabetes. Astringent, carminative, diuretic. Helps ease gas, indigestion, cramps, diarrhea.

TEA TREE • *Melaleuca alternifolia*

An evergreen tree. Essential oil is antiseptic, antibacterial, antifungal, antiviral and an immune stimulant. Eases acne, warts, cold sores, athlete's foot, yeast infections.

China and Japan

AGAR • *Gelidium amansii*

A red seaweed. Thickening agent in sauces, gelatin, jelly, ice cream. Used in science as medium for growing micro-organisms. Nutritious, fat free, lubricant, laxative.

GINGER • *Zingiber officinale*

Herbaceous perennial. Root used as a spice, antiseptic, anti-inflammatory, carminative, circulatory stimulant. Prevents motion sickness, morning sickness. Induces sweating. For sore throats, colds, coughs.

This list is solely for informational purposes.
Do not attempt to medicate yourself without the supervision of a qualified physician.

GINKGO • *Ginkgo biloba*

Ancient deciduous tree. Important use as a street tree due to high tolerance to air pollution. Enhances memory, cerebral blood flow, slows aging, antiallergenic, circulatory stimulant, anti-inflammatory, anti-asthmatic. Currently studied for its potential to treat Alzheimer's disease.

KUDZU • *Pueraria thunbergiana*

Climbing vine in bean family. Root used to treat muscle pain, measles, headache, dizziness, numbness. Edible root high in protein and starch. Invasive in the southeastern United States.

SOYBEAN • *Glycine max*

An annual bean. Important staple food with protein, iron, potassium, calcium. Said toease menopause symptoms, bone loss, and hot flashes, and reduce risk of breast cancer and heart disease. Stimulates circulation, acts as a general detoxicant.

Europe

CELERY • *Apium graveolens*

Biennial herb of carrot family. Lowers blood pressure, diuretic, antispasmodic, carminative, antirheumatic, urinary antiseptic. Seed oil used for arthritic pain. Edible vegetable.

GRAPE • *Vitis vinifera*

A perennial vine. Fruit used to make juice, raisins, and wine. Edible leaves. Has iron. Treats anemia, poor circulation, low blood pressure, skin blemishes. Astringent, anti-inflammatory. Juice eases coughs.

MULLEIN • *Verbascum thapsus*

A biennial with wooly leaves. Anti-inflammatory for colds, congestion, asthma, bronchitis and coughs. Reduces mucus. An expectorant, emollient, heals wounds and eases hemorrhoids. Flower oil helps ear infection.

OATS • *Avena sativa*

An annual grain of the grass family. Eases high blood cholesterol, stabilizes blood sugar levels. Fiber helps prevent constipation. Gentle wash for skin conditions.

PURPLE LOOSESTRIFE • *Lythrum salicaria*

An overabundant wetland perennial herb. Treats skin conditions, diarrhea, dysentery, menstrual bleeding. Antibiotic properties especially against the typhus bacterium. Astringent.

RASPBERRY • *Rubus idaeus*

A thorny perennial shrub of the rose family. Fruit used for jam and fresh fruit. Leaves ease diarrhea, menstrual cramps, and pain during childbirth. Mild astringent, diuretic.

This list is solely for informational purposes.
Do not attempt to medicate yourself without the supervision of a qualified physician.

India

Black pepper • *Piper nigrum*

Woody vine. Is aromatic, a stimulant; preserves and seasons food. Helps prevent colds, coughs, and intestinal gas. Eliminates worms and congestion. Has general warming effect. Antiseptic, antibacterial, carminative. Essential oil eases rheumatic pain and toothache.

Cinnamon • *Cinnamomum verum*

Aromatic tree, fragrant spice. Bark used as a stimulant, food preservative, antispasmodic, antiseptic, antiviral. Expectorant, decongestant, heart tonic. Eases tooth-, body-, and headaches.

Holy Basil • *Ocimum sanctum*

An aromatic annual. Lowers blood sugar and blood pressure. Antispasmodic, analgesic, reduces fever, anti-inflammatory, insect repellant, antidepressant. Stops flatulence, tonic.

Neem • *Azadirachta indica*

Evergreen tree. Antiseptic, for ulcers, eczema, as insecticide and in toothpaste. Antifungal, antiviral. Oil prevents lice and intestinal worms. Enhances immunity to malaria.

Turmeric • *Curcuma longa*

A perennial with knobby rhizomes. Eases arthritis, digestive and liver problems. Lowers cholesterol. Antibacterial, antioxidant. Helps skin conditions.

Mediterranean

Dill • *Anethum graveolens*

A carrot family annual. Used to relieve indigestion gas and cramps. Reduces bad breath. A mild diuretic, antispasmodic, used to reduce PMS symptoms. Used in pickling.

Garlic • *Allium sativum*

A biennial in the onion family. Reduces blood pressure, cholesterol, dilates vessels. A general circulatory remedy, decongestant, expectorant, helps immune function. Antifungal, antiviral, antibiotic, antiparasitic.

Lavender • *Lavandula officinalis*

Perennial shrub. A carminative, sedative. Antidepressant, antiseptic, eases cramps and tension. Stimulates blood flow. A popular fragrance for soaps and perfume.

This list is solely for informational purposes.
Do not attempt to medicate yourself without the supervision of a qualified physician.

OLIVE • *Olea europaea*
Evergreen tree. Helps digestive tract and softens skin. Reduces stretch marks. Leaves lower blood pressure and help circulation.

SAGE • *Salvia officinalis*
A mint family perennial. Used as a gargle for laryngitis, tonsilitis. Helps digestion, menstruation, menopause. An aromatic, carminative, astringent, and calms nervous system.

Unusual or Everywhere

DANDELION • *Taraxacum officinale*
A familiar weed with milky juice. Root a coffee substitute. Leaves eaten in salad. Detoxifies liver, kidney, gall bladder. Stimulates bile production. A laxative, digestive aid, and yellow dye.

COCONUT • *Cocus nucifera*
Grows on tropical beaches. Hard fruits float on ocean. Seed has protein, minerals, oil, vitamins. Used to treat burns, dissolve kidney stones, restore hair loss, soften skin, and flavor food. Component of skin and hair products.

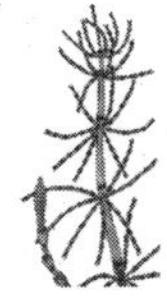

HORSETAIL • *Equisetum arvense*
Primitive plant. Stick-like leaves. Silica content helps hair, nails, eyes, eczema, acne. Clotting agent. Folk remedy for gout, poultice for wounds. Accelerates repair of connective tissues.

KELP • *Fucus vesiculosus*
A yellow or brown seaweed found near coasts. Has iodine to treat goiter. Other minerals and vitamin B12. Used in fertilizer. Laxative, immune stimulant, assists weight loss. Component of cattle feed.

LICORICE • *Glycyrrhiza glabra*
In the pea family. For coughs, colds, sore throats. Demulcent, laxative, anti-inflammatory, expectorant. Reduces breakdown of steroids in liver and kidneys. Treats liver cirrhosis and chronic hepatitis.

PEPPERMINT • *Mentha × piperita*
Hybrid annual with fuzzy leaves. Helps digestion, freshens breath, relieves gas, bloating, flatulence. Antiseptic, carminative, antispasmodic. Eases insomnia and anxiety. Used in gum and toothpaste.

This list is solely for informational purposes.
Do not attempt to medicate yourself without the supervision of a qualified physician.

Glossary

a

absorbent — The ability to take in or soak up liquids; a material or substance with that ability.
abstract — A summary of a scientific paper or research project.
abundance — The quantity of a given species within a defined area.
acid — A chemical compound that has a pH lower than 7.
acidic — Relating to or having the character of a chemical acid.
acne — A skin disorder marked by inflammation of skin glands and hair follicles and by pimple formation, especially on the face.
adhesive — A substance that sticks to another or holds two materials together by gluing, suction, grasping, or fusing.
agar — A jelly-like substance extracted from red algae and used as a culture medium.
agricultural chemist — A scientist who studies and develops chemicals used in agricultural production such as fertilizers, pesticides, and soil nutritional supplements.
agriculture — The science or practice of crop production, soil cultivation, and/or the raising of livestock.
agroecology — The study of ecological interactions in an agricultural ecosystem.
agroforestry — Sustainable growing and harvesting of forest crops.
AIDS — The acronym for acquired immunodeficiency syndrome, a disease that suppresses the functioning of the human immune system.
alkaloid — A basic and bitter organic compound, containing nitrogen and usually oxygen, found in seed plants.
alternative marketing — An industry that sell goods produced by indigenous peoples or small cooperatives in developing countries, making sure the producers receive a fair price for their efforts.
alum — Aluminum sulfate. A white, crystalline, water-soluble solid used in dye fixing, water purification, and hide tanning.
amino acid — Any of the organic acids that are the chief components of proteins and are synthesized or obtained by living cells as essential components of the diet.
anatomy — The structural makeup of a living organism.
angina — A disorder (especially of the heart) marked by spasms of intense, suffocating pain.
anthocyanin — A class of water-soluble pigments producing blue to red coloring in flowers and plants.
anthropology — The study of human cultures, relationships, and behavioral origins and development.
antiallergenic — Serving to cure or alleviate allergies.
antiarthritic — Serving to cure or alleviate arthritis.
antibacterial — Serving to kill bacteria.
anticarcinogenic — Serving to prevent or cure cancer.

antifungal — Fungus-killing.
anti-inflammatory — Serving to cure or alleviate inflammation.
antimalarial — Serving to cure malaria.
antimicrobial — Microbe-killing.
antioxidant — Serving to prevent oxidation.
antiseptic — Serving to kill or arrest the growth of microorganisms that cause decay or infection.
antispasmodic — Serving to alleviate spasms.
anxiety — An overwhelming sense of apprehension and fear often marked by physiological symptoms such as sweating, tension, and increased pulse rate.
arboretum — An institution where trees, shrubs, and herbaceous plants are grown for scientific and educational purposes.
aromatherapy — The practice of influencing moods or thoughts through the use of fragrant oils extracted from herbs, flowers, or fruits.
arthritis — Inflammation of joints.
asexual reproduction — A process by which an organism replicates itself without the union of individuals or germ cells.
astringent — Able or tending to shrink soft tissue.
atrial fibrillation — Involuntary and uncoordinated rapid twitching of muscle fibrils in the heart.

b

bacilli — A cylindrical, straight, rod-shaped bacteria.
***Bacillus thuriengensis* (Bt)** — A bacterium used as a pesticide; Bt genes have been inserted into certain crops by genetic engineers.
bacteria — A class of single-celled, simple, microscopic organisms that can be disease producers or, alternatively, valued for their chemical effects.
bar graph — A type of graph which uses parallel blocks (bars) to illustrate comparative values.
basal area — Situated at, or forming the base.
base — A chemical compound that has a pH greater than 7.
basic — Relating to or having the character of a chemical base.
bile — A bitter, greenish fluid secreted by the liver that aids the digestion of fats.
binomial — A biological species name consisting of two terms.
bioactive — Having an effect on a living organism.
bioassay — A study or experimental procedure designed to test a scientific hypothesis about an organism.
biochemical — Characterized by, produced by, or involving a chemical reaction in a living organism.
biodegradable — Capable of being broken down into organic matter by the actions of living organisms.
biodiversity — The variety of life forms in an environment.
biological illustration — A drawing that explains or clarifies the products or operations of applied biology.
biological reserve — Land set aside for the preservation of plant and animal species.
biomass — The quantity of living matter in a given unit area or volume of habitat.
bloating — Swelling by or as if by filling with water or air.
blood pressure monitor — A device that measures the pressure that blood exerts on the walls of the blood vessels.
boron — A trace element needed for plant growth, but toxic in excess.
botanical garden — An institution that contains a garden of plants and plant life; many botanical gardens also include a herbarium.

botanical sanctuaries — Privately held lands where plants can be protected from encroachment, preserved, and propagated.
botany — The study of plants or plant groups.
bowels — The intestines.

C

caffeine — A stimulating alkaloid found in coffee and tea.
calcium — An element found in teeth, bones, and many body fluids that are used in muscle contraction, nerve impulse transmission, and blood clotting.
canker sore — A small painful sore especially of the mouth.
canopy — The uppermost spreading branchy layer of a forest.
carotenoid — Similar to carotene. Yellow to red pigments widely found in plants and animal fat.
cathartic — Producing an action of purging or purification.
carbohydrate — Any of various compounds composed of carbon, hydrogen, and oxygen which are produced by plants during photosynthesis.
chemical analysis — An attempt to determine what chemicals are present in a substance or organism.
chemosynthetic — Using chemical reactions to synthesize organic compounds in order to produce energy.
chemotherapy — The use of chemicals in the treatment or control of disease, most commonly cancer.
chitin — A hardened shell or exoskeleton, especially found on insects, arachnids, and crustaceans.
chlorophyll — The green coloring in plants and leaves. Essential element in the production of carbohydrates during photosynthesis.
chlorophyll A — Chlorophyll with bright green coloring.
chlorophyll B — Chlorophyll with a dull green or khaki coloring.
chromatography — A process in which a substance is separated into its chemical components.
cholesterol — A physiologically important waxy steroid alcohol found in animal tissue.
chromatography — The separation of a mixture into its chemical components as a result of the different rates at which the components travel through or over a stationary substance.
chromosome — Any of the linear, or sometimes circular, DNA-containing bodies of viruses, bacteria, and the nucleus of higher organisms that contain all or most of an individual's genes.
cirrhosis — Hardening, especially of the liver, caused by excessive connective tissue.
climate change — A change in the average weather condition of an area.
cocci — A spherical bacterium.
companion planting — The practice of combining two types of plants in the garden in order that one may protect the other from pests.
complete protein — A protein that supplies the human body with all of the essential amino acids in the proper amounts.
complex carbohydrate — A long chain of sugar units; a carbohydrate that consists of two or more monosaccharide units.
compost — A fertilizing material consisting largely of decayed organic matter.
compress — A folded cloth or pad pressed onto a part of the body to provide pressure, moisture, or a curative substance to a wound.
concoction — A preparation made by the combining of raw materials, often including plants.
condiment — A topping or a relish used to enhance the flavor of food.
congestion — An excessive fullness or clogging, as of the heart or sinuses.

conifer — Any of an order of shrubs or trees that are usually evergreen and bear cones.
conservation — Planned management of natural resources.
constipation — Abnormally difficult or infrequent bowel movements.
contaminant — Any physical, chemical, biological, or radiological matter that damages air, water, or soil.
control — The component of an experiment that is not given or exposed to the experimental treatment or manipulated variable.
copyright — The exclusive legal right to reproduce, publish, or sell a literary, musical, or artistic work or text.
cover crop — A crop planted to prevent soil erosion and provide nutrients to the soil.
crop rotation — To rotate crops grown on a piece of land in succession over the seasons.
cultivar — A man-made variation of a species.
cultivate — To prepare and use land for the growing of crops; to foster the growth of a plant.
cutting — A stem, leaf, or root capable of growing into a new plant.
cytotoxin — Any substance that is poisonous to cells.

d

data set — A collection of related data held for analysis.
DDT (dichloro-diphenyl-trichloroethane) — A crystalline, water-insoluble substance developed in the 1950s for use as an insecticide; has toxic effects on many vertebrates.
deciduous — Falling off or out at the end of a period of growth or function, such as leaves in autumn.
decoction — An extract, usually medicinal, obtained by boiling a substance to transfer its flavor, essence, or chemical compound into the water.
decomposers — Various organisms, such as bacteria and fungi, that break down masses of organic debris into simpler substances.
deforestation — The action or process of clearing an area of forests.
defragmentation — The restoration of divided up lands into a single tract of land so as to enhance wildlife.
demulcent — A usually oily or sticky substance that soothes or protects an abraded mucous membrane.
desertification — A process in which land is made barren or desert-like, as by climate changes or mismanagement of natural resources.
detoxify — To remove a poison.
diabetes — A disorder of carbohydrate metabolism characterized by inadequate secretion or utilization of insulin, excessive urine production, excessive amounts of sugar in the blood and urine, and thirst, hunger, and weight loss.
diarrhea — Abnormally frequent and watery bowel movements.
dichotomous key — A guide to identifying plants or animals that is divided into pairs of mutually exclusive characteristics.
dietary supplement — A product that is ingested to promote the general well-being of the body or a bodily function.
disaccharide — A two-sugar molecule formed from the combination of glucose and fructose.
disinfectant — A chemical that destroys vegetative forms of harmful microorganisms but that may be less effective in destroying bacterial spores.
distribution — The natural geographic range of an organism.
diuretic — Tending to increase urine flow.
division — A propagation technique in which the roots or rhizomes of a clumped plant are divided into two or more parts to create additional plantings.
DNA — Deoxyribonucleic acid; the molecular basis of heredity, constructed of a double helix held together by hydrogen bonds.

dysentery — A disease caused by infection and characterized by diarrhea with blood and mucus in the feces.

e

ecological reserve — Land set aside for the preservation of an ecosystem.
ecological restoration — Returning damaged land to a natural and self-regulating ecosystem.
ecology — The scientific study of the relationships between organisms and their environment.
economic botany — The study of production, distribution, and consumption of plants.
ecosystem — A community of organisms (including humans) that interact with each other and with the environment in which they live.
ecotourism — Travel aimed at preserving, enjoying, and understanding nature.
eczema — An inflammation of the skin characterized by itching, oozing lesions that then become crusted.
elder — A community member having authority by virtue of age and experience.
electrophoresis — The movement of suspended particles through a fluid by an electromotive force.
emollient — Making soft or supple.
endangered — Threatened by extinction.
endemic— Something restricted to a geographic locality or region.
endorphins — A class of mood enhancing chemicals found in the body.
environmental law — A category of legal practice relating to environmental protection.
environmental journalism — A category of news that specializes in keeping the public informed about environmental issues.
enzymes— Complex proteins that are produced by living cells and aid specific biochemical reactions.
essential oil — A naturally distilled essence of a plant, used to promote health and well being.
estrogen — A substance that tends to cause sexual excitability and the development of female secondary sex characteristics.
ethnicity — Classification of people by their common traits, heritage, and customs.
ethnobiology — The study of how people of different cultures utilize living things.
ethnobotany — The study of how people of different cultures utilize plants.
ethnoecology — The study of the cultural means of management of land and ecosystems.
ethylene — A colorless flammable gas that occurs in plants as a natural growth regulator and promotes the ripening of fruit.
evergreen — Having foliage that remains green year round, as in coniferous trees.
evolution — A theory that the various kinds of plants and animals are descended from other kinds that lived in earlier times.
exotic species — A non-native species that is introduced into an ecosystem.
expectorant — An agent that promotes the discharge of mucus from the respiratory tract.
experimental design — Written instructions that describe how to conduct an experiment and what hypothesis is being tested.
experimental procedure — A defined order or series of steps in which an experiment is to be executed.
extension agent — An agricultural specialist who conducts educational outreach to assist farmers and gardeners.
extinct — No longer existing.
extract — A concentrated essence of a substance that is separated, or drawn out of a mixture or a solid.

f

fat-soluble vitamins — Vitamins that the body can store for relatively long periods of time, such as vitamins A and D.

fatty acids — The raw materials needed to make hormone-like compounds that help control blood pressure, blood clotting, inflammation, and other body processes.

fauna — Animals of a region.

fertilizer — Material for enriching land and enhancing crop growth.

fiber — Indigestible material in the human diet that stimulates the intestine to move its contents along. Strong plant material used to make clothing, paper, basketry, and other durable products.

field guide — A book containing pictures and descriptions to aid in the identification of plants and/or animals.

filial — Relating to offspring.

flatulence — Gas formed in the intestine or stomach.

flora — Plants or plant life of a region.

folklore — A culture's traditional beliefs, legends, and customs; a body of widely held, but not scientifically defined, beliefs.

forestry — The science of developing, caring for, and cultivating forests.

formaldehyde — An irritating gas used as a preservative in some manufacturing processes.

fragrant — Sweet or pleasant in smell.

frequency— A measure of the rate of occurrence of a periodic function which repeats the same sequence of values in a specific time frame.

fructose — A monosaccharide, often used as a food preservative.

fungi — Major groups of organisms that are plant-like except that they lack chlorophyll and cannot make their own food.

g

galactose — A monosaccharide formed from the breakdown of lactose and found in nerve and brain tissue.

gall stones — Calcium deposits found in the gall bladder.

gargle — To rinse the throat with liquid agitated by air forced through it from the lungs.

garland — A wreath or rope of leaves or flowers.

gel filtration chromatography — A process in which silica gel is used to separate a substance into its chemical components.

gene bank — A group of genes which are coordinately controlled; may also refer to a large collection of genetic material (e.g., seeds) intended to preserve biodiversity.

genetic base — The foundation and variety of genetic materials in a given location.

genetic diversity — A community of living creatures of a certain species, in which members of the community have variations in their chromosomes due to a number of slightly dissimilar ancestors.

genetic engineering — Artificial alteration of genetic material, including the incorporation of DNA from organism into another.

genetic resources — The available genetic base.

genetics — A branch of biology dealing with heredity and variation.

genitourinary — Relating to the genital and urinary organs.

genus — A category of biological classification that ranks between the family and species and contains related species.

GIS (geographic information system) — A computer system or software that analyzes manipulates, and displays data in a geographic context.

global environmental change — A change in the Earth's average weather conditions which in turn alters the distribution and abundance of organisms on the planet.

global warming — The abnormally rapid heating up of the atmosphere, caused by the release of carbon dioxide, carbon monoxide, and other gases.
glucose — A simple sugar found in blood; the body's main source of energy.
goiter — An abnormally enlarged thyroid gland visible as a swelling at the base of the neck.
gout — A metabolic disease marked by painful inflammation and swelling of the joints.
Gram-negative — The result of Gram's method of bacteria staining in which the bacteria tested have a cell wall surrounded with an extra layer of lipopolysaccharides that do not retain violet dye.
Gram-positive — The result of Gram's method of bacteria staining in which the bacteria tested have a thick cell wall made of disaccharides and amino acids that retains the violet dye.
green manure — An herbaceous crop plowed under while green to enrich the soil.
Gross National Product — A measure of the total value of the goods and services produced by the citizens of a country during a specified period.

h

habit sketch — A preliminary outline drawing of an entire plant used as a basis for a more formal or detailed drawing.
habitat — A location or environment where a plant or animal naturally or normally lives and grows.
habitat fragmentation — The reduction of a given habitat type in a landscape and/or the division of remaining habitat into smaller, more isolated parcels.
habitat loss — The destruction or removal of the location or environment where a plant or animal naturally or normally lives and grows.
heirloom — Something handed on from one generation to another.
herb — A seed plant that lacks woody tissue and dies to the ground at the end of the growing season.
herbaceous — Lacking woody tissue.
herbal — Written collections of accumulated plant knowledge and lore, often augmented with the observations and experiences of the author.
herbarium — A collection of dried plant specimens.
hemorrhoid — A swollen mass of dilated veins at the anus.
histogram — A graphical representation of frequency distribution.
hormones — Products of living cells that circulate in body fluids and have a specific effect on the activity of cells remote from their point of origin.
horticultural therapy — A form of therapy that uses the activity of cultivating live plants to heal and rehabilitate people.
horticulture — The science and art of growing fruits, vegetables, flowers, and ornamental plants.
hyphae — Thread-like filaments that make up mycelium in most fungi.
hypothesis — An assumption made in order to test its logical or empirical consequences; an educated guess.

i

immunization — Administering a medicinal agent in order to render a high degree of resistance to a disease.
incense — Material used to produce a fragrant odor when burned.
increment borer — A tool used to take out a core from a tree to determine its age and health.
incubate — To maintain a favorable environment to promote the development of an organism.

indigenous — Produced, growing, or living naturally in a particular region, as a native species or population.
informant — Someone with firsthand experience who can provide social or cultural information.
infused oil — A distilled oil created by soaking or steeping plant tissue in liquid that then obtains some of the plant's properties or ingredients.
infusion — A flavored or medicinal beverage made from leaves or leaf buds soaked in water.
indigestion — Inadequate or difficult digestion.
inoculate — To introduce a microorganism into an organism, especially in order to treat or prevent a disease.
insecticide — A preparation for destroying insect pests.
insect repellent — A substance used to ward off insects.
insomnia — An inability to sleep.
intellectual property rights — The notion that a person who develops a body of unique knowledge or an idea should have some control over how that knowledge or idea is used and should receive a fair share of the profits that come from it.
internship — A program whereby an advanced student gains supervised practical experience in a professional field.
invasive species — An aggressive species that can take over the habitat of others.
invertebrate — A living organism that lacks a spinal column.

j

jargon — The technical terminology or characteristic language of a particular profession or hobby.

k

key — *See* dichotomous key.
keystone species — A species upon which many others depend.
kidney stones — Calcium deposits found in the kidney.
kingdom — A biological category that ranks above the phylum.

l

land race — A locally used variety of a crop, usually developed by farmers for their own use.
landscape design — The art of arranging or modifying the features of a natural landscape for aesthetic reasons.
land-use change — An alteration of the way people use a piece of land, often resulting in a permanent change in the flora of the area.
lanolin — A fatty substance extracted from wool which is used as a lubricant in cosmetics, ointments, and soaps.
laryngitis — Inflammation of the larynx, often causing a temporary loss of voice.
latex — A milky juice produced by the cells of certain plants.
laxative — A preparation that relieves constipation.
layering — A propagation technique in which a shoot or stem is bent into the ground to root while still attached to the living plant.
light meter — An instrument used to measure the intensity of light on a certain site.
line graph — A graphical representation using vertical lines of varied heights to plot data that describes changes that take place over time, or to describe a relationship between two continuous variables.
linguistics — The study of human speech and languages.

linolic acid — A fatty acid found in some plants.
lipids — A group of fats that are stored in the body as the most concentrated sources of food energy.
literature review — The stage of research during which one reads books and journal articles relevant to one's topic of interest, to review what other scientists have thought and discovered.
litmus paper — A type of specially treated colored paper that is used to measure pH.
lore — A body of traditional knowledge on a particular subject.
lubricant — A material or substance capable of reducing friction.
lumbago — Rheumatic pain in the lower back and loins.

m

mace — A spice made from the fibrous coating of nutmeg.
macroelements — Elements found in large amounts.
macronutrient — Required nutritional elements of a diet which are required in large amounts, including carbohydrates, fats, and protein.
magnesium — A mineral element that is essential for many body functions, including nerve impulse transmission, the formation of bones and teeth, and muscle contraction.
malaria — A disease marked by recurring chills and fever; caused by a protozoan parasite of the blood transmitted by the anapholes mosquito.
manipulated variable — A variable that is managed or controlled in an experiment.
mapping software — A computer program that helps create maps that indicate elevation, landforms, and a variety of other data.
market value — The current monetary value of something, in terms of how much it may be sold for fairly in an open market.
mean — The average of a set of numbers.
median — The middle term in a set of numbers arranged in order from lowest to highest.
medicinal — Used to cure disease or reduce pain.
menopause — The period in a female primate's life when menstruation stops naturally.
menstruation — Discharging of blood, secretions, and tissue debris from the uterus at approximately monthly intervals of breeding-age nonpregnant primate females.
metabolism — The chemical processes in living cells whereby energy is created for the organism.
microbial agent — Microscopic organisms that may cause illness, including all potentially dangerous species of fungi, bacteria, viruses, or protozoa.
microbe — A microorganism.
microelements — Elements found in small amounts.
micronutrient — Required nutritional elements of a diet which are required in small amounts; these include trace minerals and vitamins.
microorganisms — Organisms that cannot be seen without the aid of a microscope.
micropropagation — Starting plants in test tubes using only a few cells to grow new plants.
mineral — In nutrition, any crystalline, inorganic nutrient that is essential in small quantities for body structure and the proper functioning of body processes.
mode — The most frequently occurring value in a set of data.
modernize — To give new or modern characteristics to an old object or practice.
moisture meter — A device that measures diffused or condensed liquid in or on the surface of an object.
monosaccharides — Simple, single sugars with carbon, hydrogen, and oxygen in a 1:2:1 ratio.
mordant — Biting, pungent.
morphology — A branch of biology that deals with the form and structure of an organism or any of its parts.

motion sickness — Nausea induced by motion or physical disorientation.
mucus — A slimy, slippery protective secretion of membranes.
mulch — A protective covering spread over the ground to reduce evaporation or weeds.
mutation — A significant and basic alteration, as of a gene.
mycorrhizae— The symbiotic relationship of a fungus with the roots of a plant.

n

native — Grown, produced, or originated in a particular place.
natural history — The study of the natural development of something over a period of time.
naturalist — Someone who educates people about the natural world and the environment, often in nature preserves, camps, and state and national parks.
natural resources — The wildlife, forests, mineral deposits, oil, and other natural materials found in a geographic area which may be exploited by industry.
nausea — Sickness of the stomach with a desire to vomit.
nitrates — A class of nitrogen-based chemicals used as fertilizer.
non-native species — A plant species introduced into an area where it did not originate.
non-woody — Herbaceous plants having little or no woody tissue.
noninfectious — Diseases originating from behavioral, genetic, or environmental factors, rather than spreading by pathogens.
nutrient — A substance that is absorbed by a living organism that requires that substance for growth, energy, or the sustenance of life.

o

ointment — An oil- or grease-based preparation applied to the skin, usually for medical purposes.
organic — Derived from living or formerly living things.
organic matter — A substance that contains carbon; materials that exist in, or are derived from, living things.
ornamental — Something that lends beauty.

p

parasite — A plant or animal living in, with, or on another living organism, usually causing harm to the host.
pathogen — A disease-causing agent.
pest — An insect, rodent, fungus, weed, or other life form that is harmful to one's health or the environment.
pesticide — A natural or synthetic chemical agent used to destroy insects or small animals that are harmful to an environment.
petri dish — A glass or plastic dish used to grow and study microorganisms in culture.
pH — A measure of acidity or alkalinity.
pharmaceutical — Engaged or related to pharmacy.
pharmaceutical drugs — Medicinal compounds used to treat specific disorders.
pharmaceutical prospecting — Searching for bioactive compounds with a potential for development as commercial drugs.
pharmacologist — A person engaged in selling or making medicinal drugs.
photosynthesis — The process by which chlorophyll-containing plants use carbon dioxide, water, and the energy of the sun to produce carbohydrates and oxygen.
physiology — A branch of biology dealing with the functions and the functioning of living matter and organisms.

phytochemical — Any of a group of chemical compounds in plants and plant products.
phytoremediation — Utilizing plant materials to clean up toxic wastes.
pie chart — A graphical means of representing quantities as a circle divided into parts that represent values by their relative sizes.
placebo — An inert medication used for its psychological effect.
plagiarism — Presenting someone else's words, images, or ideas as one's own; failure to give proper credit to someone whose material one uses.
plant biochemistry — A branch of science that studies medicinal compounds, the nutritional value of plants, and plants as sources of useful natural products such as perfumes, resins, tannins, or oils.
plant breeder — A botanist or horticulturist who develops new varieties of crops or ornamentals.
plant diversity — Variation of plant species in a given region.
plant press — A machine used to extract moisture from plants for preservation.
plant taxonomy — That branch of botany that identifies, classifies, and names plants according to an orderly system.
pollen — A mass of male spores of a seed plant.
pollutants — Agents that contaminate an object or area.
polyculture — The practice of growing and raising more than one species of plant in the same environment at the same time.
polypeptide — A short chain of amino acids produced as the result of protein digestion.
polysaccharide — A complex sugar or starch.
population — A group of individual persons, organisms, or items from which responses or samples are taken for statistical measurement.
potassium — A mineral element that helps the human body maintain water balance, normal heart rhythm, conduction of nerve impulses, and muscle contraction.
potpourri — A mixture of flowers, herbs, and/or spices used for scent.
poultice — A soft, heated mass of natural medicinal substances, such as herbs, spread on cloth and applied to an area of skin for medicinal purposes.
prescription — A written direction for a therapeutic or corrective agent, typically issued by a physician.
primary producers — Green plants, which transform solar energy into matter through the process of photosynthesis.
propagate — Produce or cause to reproduce biologically.
prostratin — An ingredient found in the inner bark of a *mamala* tree which, *in vitro,* helps protect healthy cells from infection by the HIV-1 virus.
protein — Any of a group of organic chemical compounds that are essential to the structure and function of living things.
protist — A microorganism of the Protista kingdom.
protocol — A set of conventions governing the correct procedures of a scientific experiment, treatment, or procedure.
protozoan — Any of a phylum of unicellular lower invertebrate animal like protist.
pseudo-medical — Language that is used with the intent to sound official or medically knowledgeable, often used to mask a lack of scientific basis.
pulp — A soft mass of shredded vegetable matter from which most of the water has been extracted by pressure.
purgative — A strong laxative.

q

quadrat — Any of the four quarters into which something is divided by two lines intersecting each other at right angles.

quarantine — A state of forced isolation, most often used to stop or prevent the spread of disease.

quinine — A bitter white alkaloid obtained from cinchona bark and used in producing quinine-based seltzer drinks and as a remedy against malaria.

r

radiation — To send out rays.

rain forest — A tropical woodland having an annual rainfall of at least 100 inches.

range — Open land where animals may roam and graze.

recombinant DNA — Genetically engineered DNA prepared by splicing together specific DNA fragments usually from more than one species of organism.

recycle — To process in order to regain and reuse material for human consumption.

regulations — An authoritative set of rules that deal with details or procedure.

reliability — In science, a measure of the extent to which an experiment may yield the same results on repeated trials.

resin — Any of various substances obtained from the gum or sap of some trees.

responding variable — In an experiment, the change that is measured.

response — The activity of an organism or any of its parts resulting from stimulation.

restoration ecology — That branch of ecology that attempts to return damaged land to a natural and self-regulating ecosystem.

rheumatism — Any of various conditions marked by stiffness, pain, or swelling in muscles or joints.

RNA — Ribonucleic acid; nucleic acids that are associated with the control of chemical activities on a cellular level.

s

sachet — A small bag filled with perfumed powder for scenting clothes.

salves — Medicinal substances applied to the skin.

saprophytes — Plants that derive nutrition from dead or decaying organic matter.

saturated fat — A type of edible fat, solid at room temperature, that increases the cholesterol level in human blood.

scarification — A process of cutting the walls of a seed in order to hasten its germination.

sciatica — Pain in the region of the hips.

sedative — Tending to relieve tension.

seed bank — A storage facility to preserve seeds.

serotonin — A chemical in the brain that transmits nerve impulses, causes blood vessels to constrict at sites of bleeding, and stimulates muscle movement in the intestines; also believed to play a role in mood.

sexual reproduction — The process where two cells fuse to form one hybrid, fertilized cell offspring whose DNA is different from that of either parent.

shaman — A wise or holy person who is believed to have knowledge of the application of resources from the natural or supernatural world.

shelf life — The period of storage time during which a material will remain useful or edible.

silica gel — An absorbent used in gel filtration chromatography.

species — A category of biological classification ranking just below genus or subgenus.

spectrophotometry — The measurement of visible electromagnetic radiation.

spectroscopy — An instrument to see visible electromagnetic radiation on film or scope.

spermicide — A preparation used to kill sperm.

spice — Any of various aromatic plant products.

spirilla — Long curved bacteria, usually having tufts of flagella.

standard deviation — A measure of dispersion in a set of data.
statistical analysis — Close examination of the numerical data gathered in any given experiment.
starch — A carbohydrate that the body digests, absorbs into the bloodstream, and uses for energy.
sterile — Incapable of producing offspring, fruit, or spores; a device or instrument cleaned so as to be incapable of transmitting microbial agents.
steroids — Any of numerous compounds including various hormones and sugar derivatives.
steward — One who manages or supervises the appropriate use of land, property, money, or other resources entrusted to his or her care.
stewardship — Protecting and caring for the land and the environment.
stimulant — An agent that temporarily increases the activity of an organism or any of its parts.
subject — The object of research whose reactions or responses are studied.
succulent — Any of the various plants that have fleshy tissues that conserve moisture, such as cactus.
sugar — Any of several sweet, crystalline, water-soluble substances such as glucose, lactose, and fructose.
sustainable — Able to continue to thrive and provide nourishment over a long period of time.
symbiotic relationship — A mutually beneficial relationship between two organisms.
synthetic pesticide — A man-made combination of chemicals that is used to destroy insects or small animals that are considered harmful or undesirable.

t

t-test — A statistical test used to determine the limits for the random variable t of a distribution. It is used to test hypotheses about means of normal distributions when the standard deviations are unknown.
taxonomy — Classification of plants and animals according to natural relationships.
tea — A flavored or medicinal beverage made from dried leaves or leaf buds steeped in boiling water.
textile — Any of various fabrics, usually woven and often made of natural materials such as cotton.
textural triangle — A diagram used to help determine the texture of a soil sample.
threatened — In danger of becoming extinct.
tincture — An alcoholic solution of a medicinal substance.
tonic — Something that invigorates, restores, or refreshes.
tonsillitis — Inflammation of the tonsils.
toxicity — The amount of poison in a substance.
toxins — Poisonous substances that are notably hazardous when introduced to living tissues, and typically capable of inducing antibody formation.
transect — A sample area of vegetation, usually in a long strip.
transpiration — The process by which living plants release water vapor into the atmosphere.
treatment — A remedy or series of measures designed either to relieve or cure a disease or ailment or to induce a response in an experimental subject.
triglycerides — An important energy source in the body. The main form of fat in the blood.
tropical — Of the geographic region situated between the Tropic of Cancer (23.5 degrees North latitude) and the Tropic of Capricorn (23.5 degrees South latitude).
tropical rain forest — A tropical woodland with an annual rainfall of at least 100 inches and marked by lofty broad-leaved evergreen trees which generally form a continuous canopy.

tuber — A short, fleshy, underground stem.
type specimen — The herbarium specimen on which a description of a new plant species is based.
typhus — A severe infectious disease transmitted by body lice, caused by a rickettsia bacterium.

u

ulcer — An open eroded sore of skin or mucus membrane.
United States Pharmacopeia (USP) — The agency that sets and maintains standards of quality for natural medicines and nutritional suppliments.
unsaturated fat — A type of fat, liquid at room temperature, that does not seem to increase cholesterol level in blood.

v

variety — A category of biological classification ranking just below species or subspecies.
vegetative propagation — Asexual reproduction through growth of a plant part.
vertebrates— Any animal possessing a segmented spinal column.
virus — Any of a large group of submicroscopic infectious agents that have an outside coat of protein around a core of RNA or DNA.
vitamin — Any of various organic substances that are essential in tiny amounts to the nutrition of most animals and some plants.
volatile oil — A distilled oil, generally obtained from plant tissue, that evaporates or dissolves rapidly.
voucher specimen — An herbarium specimen on which a study is based.

w

warts — A small, tough, projecting growth on the skin.
water-soluble vitamin — A vitamin that is not stored in the body in significant amounts and needs to be replenished on a regular basis, such as C and the B-complex vitamins.

x

xanthophyll — A class of water-soluble pigments producing yellow coloring in flowers and plants.
xeriscaping — A form of landscaping and gardening where plants are selected which are suited to the climate and which uses water sparingly.

y

yeast — A minute one-celled fungus that reproduces by budding.

z

zinc — An essential trace mineral that is involved in over 20 different enzymatic reactions in the human body.

Index

p

q

r

s

t

u

w

z

Gabriell DeBear Paye grew up in New York City. She earned a B.S. in Environmental Horticulture with a minor in Botany in 1983 and an M.A. in Education in 1984 from the University of Connecticut. From 1985 to 1987 she served as a Peace Corps Volunteer in Liberia, West Africa, where she taught biology and agriculture and learned firsthand about tropical agriculture and ecology. In 1987 she came to Boston to teach biology and horticulture at West Roxbury High School. Following an Earthwatch trip to Mexico in 1994 to work with ethnobotanist Bruce Benz, she began writing and testing ethnobotanical activities in her classroom. In 1994 she launched an ethnobotanical web site for teachers on the EnviroNet network and in 1996 she was selected as an Access Excellence Fellow with Genentech. She received a Global Teachnet award from the National Peace Corps Association in 1997 for her ethnobotanical curriculum materials. In 1999 she was awarded a Pioneer Grant with MetroLinc and a Teachnet Grant through Impact II for her botanical curriculum materials. She is a Lead Teacher in the Boston Public Schools, and is the mother to her daughter, Amity.